Excel® Student Laboratory Manual and Workbook

Johanna Halsey
Dutchess Community College

Ellena Reda
Dutchess Community College

The Triola Statistics Series:

Elementary Statistics, Eleventh Edition

Elementary Statistics Using Excel, Fourth Edition

Essentials of Statistics, Fourth Edition

Elementary Statistics Using the Graphing Calculator, Third Edition

Mario F. Triola
Dutchess Community College

Addison-Wesley
is an imprint of

PEARSON

Reproduced by Pearson Addison-Wesley from electronic files supplied by the author.

ISBN-13: 978-0-321-57073-4
ISBN-10: 0-321-57073-1

3 4 5 6 BRR 11 10

Addison-Wesley
is an imprint of

www.pearsonhighered.com

Preface

It is important to note that this manual was written to support the use of Excel 2007. Previous versions of Excel are significantly different from the 2007 version. If you are using an earlier version of Excel you may want to ask for an earlier version of this manual.

Purpose of the Manual

This manual contains step-by-step instructions to help you familiarize yourself with the spreadsheet program **Excel 2007**. Our primary purpose in creating these instructions is to help you become proficient with those Excel features that support working with data in your statistics class.

We know it is impossible to cover every possible option in using the program, so we have chosen the techniques that have worked well for us and our own students. There are usually at least two ways to produce the same results. Our hope is that providing you with a solid set of step by step instructions, you will become comfortable enough with the program to begin to experiment on your own, and share your discoveries with other students in your class.

Layout of the Manual

Other than Chapter 1, this manual follows *Elementary Statistics, 11*[th] Edition section by section. The exercises worked through in the tutorial instructions are based on the presentation of the material in the corresponding section in your text. At the end of each section you will find reference to several exercises which will give you an opportunity to practice the technology skills introduced within that section. Our hope is that you will immediately employ the features you learn in **Excel 2007** to help you with the exercises and projects presented in that section of your textbook. You will find that using the technology to work on many exercises in the book will afford you the practice necessary to become proficient with the program.

Using Technology Wisely

Any software package has its own learning curve. You should expect it to take a certain investment of time, energy and regular practice to become comfortable with the program, particularly with this version of Excel. The more regularly you commit to using the ideas presented in this manual, the more proficient and adept you will become with using this program when and where appropriate.

While there are many, many places that Excel can, and should be integrated into the course material, you also can benefit from doing some of the work without using any technology. To truly understand some concepts, you need to perform at least some of the computations by hand. Once you have a core understanding of an idea, the technology affords you a way to find answers quickly and accurately.

Technology Notes

The instructions contained in this manual are written for a PC. The MAC version of Excel 2008 is significantly different from the Windows version of Excel 2007. The instructions in this manual are not compatible with MACs.

Early in the manual, you will be asked to load the Data Desk®/XL (DDXL) Add-In that accompanies your textbook. This Add-In supplements Excel, providing additional statistical tools not included in Excel. For example you will use DDXL to construct boxplots, confidence intervals and to perform hypothesis tests.

You can find the data sets from Appendix B of your textbook on the CD-ROM that accompanies your text. You can also download these data sets from the Internet at http://www.aw-bc.com/triola.

Final Notes

We feel the benefits of using Excel in a statistics course are vast. Many companies look for employees who are proficient with using spreadsheets. By learning how you can use Excel to support your work in statistics, you will simultaneously be developing a skill that is highly valued in the business world. Our hope is that this manual provides you with a relatively painless entry into the world of spreadsheets! We hope you enjoy your learning journey.

Johanna Halsey

Ellena Reda

CONTENTS

CHAPTER 1: GETTING STARTED WITH MICROSOFT EXCEL

SECTION 1-1: INTRODUCTION

It is important to note that this manual was written to support the use of Excel 2007. Earlier versions of Excel are significantly different from the 2007 version. If you are using an earlier version of Excel you may want to ask for an earlier version of this manual.

One of the most valuable computer programs used in business today is the spreadsheet. A spreadsheet program is an electronic workbook that allows you to enter, edit, and change the appearance of data. It allows you to perform calculations, organize and analyze data, and show relationships in data through various types of charts and graphs. Data can be entered and stored in Excel as text, numbers or formulas.

There are many practical applications for using a spreadsheet. For example, industry and business may use Excel for finance, to analyze market trends, and for forecasting. Your instructor may use Excel to record and determine your course grades. Many computers come preloaded with Microsoft Office®. Excel is part of the Microsoft Office Suite. It is a popular spreadsheet program and can be used by all skill levels. Excel has the capability to perform tedious operations on very large sets of data quickly.

The purpose of this chapter is to provide you with an introduction to spreadsheets and to prepare you to use Excel in the study of statistics. If you have not had any experience with a spreadsheet program this chapter will be invaluable to you. We suggest that you take your time and work through the step-by-step instructions in this chapter. You will be glad that you did.

The best way to become familiar with most software packages is to dive in. As you explore Excel you will notice that there are often several ways to perform the same task. This manual will highlight only one or two of those ways. However that should not prevent you from trying other ways or to uses a method you are already familiar with.

A Special Note to Mac users: The instructions contained in this manual are written for a PC. The MAC version of Excel 2008 is significantly different from the Windows version of Excel 2007, so the instructions in this manual may be of limited use to MAC users.

SECTION 1-2: THE BASICS

If you have worked with earlier versions of Excel you will immediately notice that Excel 2007 looks very different from those earlier versions. All of the same features of those earlier versions are contained in this version of Excel. The new redesign of Excel is meant to make your work easier. As part of the redesign of Excel 2007 Microsoft introduces three new features: the Microsoft Office Button, the Ribbon, and the Quick Access Toolbar. We will explore each of these features in the following pages.

You may open the Excel program one of two ways:

1) If you have an Excel icon on your desktop you can **double click on the Excel icon** to open the program. This is generally the easiest way to access Excel.

OR

2) If you do not have an Excel icon on your desktop you can access Excel as follows:

a. **Place your cursor** over the word **Start** located in the lower left hand side of your computer screen.

b. **Click on your left mouse button** to access the **Start menu.**

c. Move your cursor so that you **highlight All Programs**

d. Move your cursor to the right. Locate and **highlight Microsoft Office.** You will do this by moving your cursor up or down the list of programs shown until you have found Microsoft Office.

e. Move your mouse to the right. Locate and **highlight Microsoft Excel 2007**

f. Click on the left mouse button (Mac users just click).

When you start Excel, a blank worksheet appears as seen below. This worksheet is the document that Excel uses for storing and manipulating data.

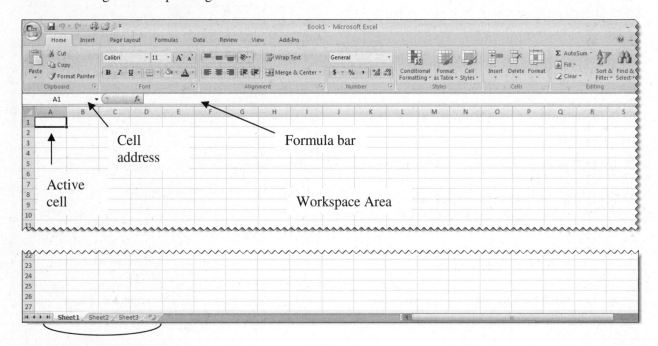

Cell address

Formula bar

Active cell

Workspace Area

Sheet tabs

The following are some **basic terms** that you should be familiar with. Try to locate each of these on your Excel screen. Let's start with the new features of Excel 2007: the Ribbon, the Microsoft Office Button, and the Quick Access Toolbar.

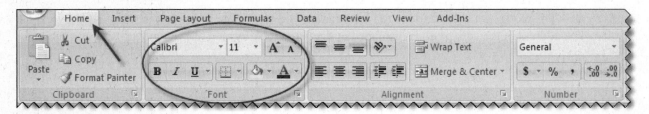

The **Ribbon** is the panel on the top portion of the document and contains commands that are grouped together by task. There a seven **tabs** on the Ribbon. The **Home tab** is identified in the Ribbon on the preceding page. Each tab is divided into **groups** that represent a collection of features that are related. Below each group is a name that identifies the tasks associated with that group. The features circled in the Ribbon on the preceding page all focus on formatting fonts.

The **Microsoft Office Button** is found in the upper left hand corner of your screen and contains common file and system commands. It contains many of the functions that were located under the File menu in older versions of Excel. When you click on the left mouse button a larger menu will be available to you. The Microsoft Office Button allows you to open a new or existing workbook, save files, print your work and close files.

The **Quick Access Toolbar** is found in the upper left hand corner of the screen. It is highlighted by a box in the figure shown here. This toolbar contains shortcuts for some of the more commonly used commands.

You can customize the Quick Access Toolbar by adding the commands that you most frequently use.

1) **Left click on the down arrow** to the right of the Quick Access toolbar.

2) **Highlight the item** you wish to add to the Quick Access Toolbar.

3) **Left click** on that item to add to the toolbar.

4) If you are looking for a command that does not appear on this list choose **More Commands**.

5) Select the command you wish to add and then left click on the **Add** button.

6) Left click the **OK** button when you are finished.

Cell – located at the intersection of a row and a column. Information is inserted into a cell by clicking on cell and entering the information directly.

Cell address – Location of a cell based on the intersection of a row and a column. In a cell address the column is always listed first and the row second so that A1 means column A row 1.

Active cell – The worksheet cell receiving the information you type. The active cell is surrounded by a thick border. The address of the active cell is displayed in the Name box located above the worksheet on the left.

Formula bar – Area near the top of the Excel screen where you enter and edit data.

Sheet tabs – Identify the names of individual worksheets.

Worksheet area – The grid of rows and columns into which you enter text, numbers and formulas.

SECTION 1-3: ENTERING AND EDITING DATA INTO EXCEL

When an Excel worksheet is first opened, the cell A1 is automatically the active cell. In the screen on the right you will notice that A1 is surrounded by a dark black box. This indicates that it is the **active cell**. Note that A1 is also shown in the **Name box** as well.

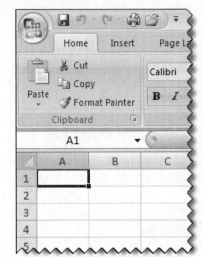

If you press a key by mistake, accidentally delete information you may have wanted or find the screen isn't doing quite what you expect it to do try left clicking the **Undo button** located in the Quick Access toolbar. It looks like an arrow looping to the left.

THE FORMULA BAR

The formula bar is located just below the Ribbon in your Excel worksheet and to the right of the Name box. The first window on the formula bar shown below indicates the cell address of the active cell. The **X** is used when we wish to delete information we have typed into the active cell. The **check mark** takes the place of the Enter key. Click the left mouse button to indicate that the data or formula you have entered into the active cell is acceptable. You can also just use the **Enter key** on your computer keyboard.

There are three types of information that may be entered into a cell:
- Text
- Numbers
- Formulas

ENTERING TEXT AND NUMBERS

The following exercise will give you some practice in entering both text and numbers into a worksheet.

1) **Position the cursor** (sometimes called the insertion point) on cell A1 and click on the left mouse button. This makes cell A1 the active cell.

2) **Type** MONTHLY EXPENSES in cell A1 and press **Enter.** Notice that the active cell is now A2.

   a. In cells A2 through A7 type Rent, Food, Utilities, Cable, Telephone and Internet. Press the **Enter** key or the **down arrow** key to move down to the next cell.

   b. Follow the same process to enter the amount spent each month for these categories. These amounts will be entered in cells B2 through B7.

Don't worry if the dollar signs do not appear when you enter the dollar amounts. We will look at how to format these cells next.

Note:

To move to the cell just below a cell you have entered data in press **Enter.**

To move to the right of a cell you have just entered data in press **Tab.**

You can also move in any direction by using the **arrow keys** on your computer.

FORMATTING CELLS

The information that you just entered into the Excel spreadsheet in the preceding example represents dollar amounts, although this may not be apparent when you first enter the data.   Many of the commonly used formatting tools are located on the **Home Ribbon** and are grouped according to their related tasks. The **Number group** circled below shows the formatting icons that you will use to format numbers in your Excel worksheet.

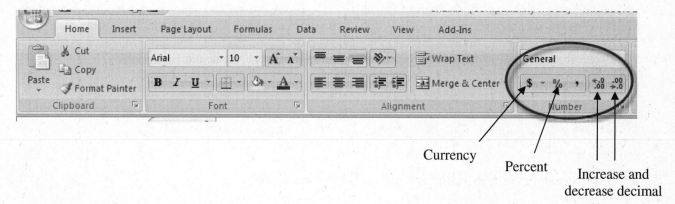

Currency   Percent   Increase and decrease decimal

The **increase decimal or decrease decimal** icons both allow us to determine the number of values that will be displayed to the right of the **decimal.**

To Format Cells

There are two different ways to change the values you have entered in your Excel spreadsheet so that they will represent dollar amounts. These methods are outlined here.

Method 1:

1) **Click on** the **B** at the top of the second column. This allows us to select the entire column rather than individual cells. The entire column is now highlighted.

2) Click on the currency icon ($) found in the Number group in the Home tab.

3) All data in that column should now contain a dollar sign in front of it.

Method 2:

1) **Select the cells** you wish to format.

2) Place your cursor over the highlighted area.

3) Click on your right mouse button.

4) Scroll down to **Format Cells** and left click.

5) This will open a **Format Cells** dialog box. This dialog box presents a number of options to you, that include the ability to choose the type of number and the number of decimal places you wish displayed.

Try using this method to change all of your data in your worksheet to dollar amounts at the same time.

SAVING YOUR WORK

Once you have entered data into an Excel spreadsheet you will want to save your work. This can be done by either left clicking the **Save** icon on **Quick Access Toolbar** (it looks like a floppy disk) or by left clicking on **Microsoft Office Button** and choosing the **Save As** command. You will use the Save As command the first time you save and name a file. In either case a dialog box will appear similar to the one you see below.

a. Choose **Desktop** to save the information to your desktop.
b. Choose **C:/My Documents** to save information to your **hard drive.**
c. If you use an external flash drive choose the drive location (in this case we would choose the Lexar Media (E:) to save our file).
d. Enter a **file name** that will accurately indicate what the file contains and then left click on **Save**.

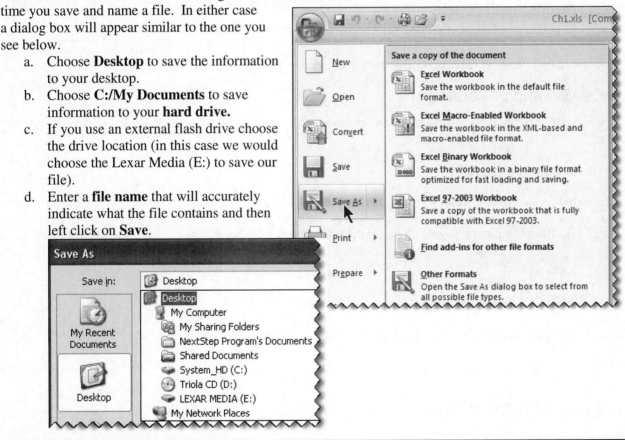

Note:
It is a good idea to follow the wise old adage of "save and save often". It can be very frustrating to spend a good amount of time working on something only to lose your work due to a computer glitch. Be mindful that when working on a spreadsheet, it is always a good idea to click on the **Save** icon periodically.

EDITING INFORMATION

After data has been entered you can edit the information in a cell by **activating** that cell (left click on it). The information in that cell is now displayed in the formula bar. Move your cursor to the entry you wish to modify or change. The cursor will turn into what looks like the capital letter I (remember that this is commonly referred to as the insertion point). Place the I-beam at the point you wish to make changes, left click and proceed from there.

SELECT A RANGE OF CELLS

There are many times when working in Excel that it is useful to select more than one cell at a time. A group of selected cells is called a **Range** of cells.

To select more than one cell at a time

1) **Left click on a cell** (in this case we have activated cell B3) and hold the left mouse button down.

2) With the cursor in the middle of the cell move down to cell B8.

3) **Release the left mouse button.** The cells B3 through B8 should now be highlighted.

We refer to this **range of cells** as **B3:B8**, using a colon to separate the first cell in the range from the last cell.

DROPPING AND DRAGGING

To easily move a range of cells from your worksheet we can use a shortcut method known as **drop and drag**.

To move a range of cells

1) Select the range of cells you wish to move using the method outlined above.

2) **Position the cursor** anywhere on the cell range border so that it changes to a move pointer which looks like this.

3) **Hold** the left mouse key down.

4) **Drag the range of cells** to their new location.

5) **Release** the left mouse button.

COPYING A RANGE OF CELLS

To copy a range of cells

1) Select the range of cells as you did above.

2) After you release the mouse button **left click** on the **Copy icon** found in the **Clipboard group** in the **Home** tab.

3) **Activate** the cell that you want to copy the data to. Recall that we activate a cell by simply using a left click on the cell we wish to select.

Copy icon

4) **Left click** on the **Paste icon** found to the left of the copy icon as seen here.

OPENING FILES IN EXCEL

As you work through this some of the examples found in this manual and on exercises from your textbook, you will need to open files you have saved to a flash drive or to your hard drive. You will also be asked to use data sets found on the CD-ROM Data Disk that came with your Statistics textbook. These data sets are the same as the data found in Appendix B of your text. You can also find the data sets for the textbook at the website www.aw-bc.com/triola.

To use the data sets on the CD-ROM

1) Open Excel and place the CD into the CD-ROM drive on your computer. On most computers this will show up as the D drive.

2) **Left click** on the **Microsoft Office Button** located in the upper left hand corner of your workbook.

3) Select **Open** from the menu that is presented to you. **Left click** on the **Open** icon.

4) **A dialog box** similar to the one we saw when we discussed saving your work will open. It will look similar to the one shown here. Notice that the CD for this textbook is located in the D drive. The title of the CD is shown as well.

5) **Left click on Triola CD.**   This will open a second dialog box. This shows several folders. You will want to choose the folder titled **App B Data Sets**. This folder contains the same data sets found in Appendix B of your textbook.

6) **Double click** on this folder. This presents another dialog box which is shown on the following page. In this dialog box choose the folder titled **Excel** for a complete list of the files already in Excel workbooks.

7) Locate the file you are looking for by file name. **Left click** on the **file name**.

8) **Double click on the file name** or **left click** on **Open** found on the lower right hand corner of the dialog box.

Opening your own files in Excel

You will follow the same process if you are trying to open a file that contains data that you have saved to your desktop, hard drive or a flash drive. Follow all of the steps in the same sequence beginning with Step 2. In Step 5 rather than selecting the D drive you would choose the location that contains the file that you are trying to open. Choose **Desktop** to retrieve the information saved to your desktop. Choose **C:/My Documents** to retrieve any files you may have saved to your **hard drive.** If you use an external flash drive choose the drive location identified by your computer to retrieve a file save to the flash drive.

Downloading files from the textbook website

As previously mentioned you can also find the same data sets found in your textbook at the companion website www.aw-bc.com/triola.

To download these files to your computer go to this website. Choose the textbook that corresponds to the one you are using in your statistics class. Choose Appendix B Data Sets. Follow the directions found on the website to download these files to your computer. Be sure to pay attention to where the files are saved on your computer so that you can find them easily.

SECTION 1-4: UNDERSTANDING AND USING FORMULAS:

When we analyze data in Excel it is often necessary to use formulas which are equations that perform calculations on the values you have entered into your spreadsheet. When you enter a formula into Excel it is important that you always place an equal sign (=) at the start of your formula. This is how you communicate to the software that you are entering a formula rather than data or text.

When we enter a formula in Excel we can either use the actual value found within a cell or the cell address. As you you're your formula it will be displayed in the formula bar. After you have typed in the formula and pressed the Enter key, Excel automatically performs the calculation and displays the numerical result in the cell you have chosen. If you change a number stored in a cell address used in the formula, Excel automatically recalculates the results of that formula. You can also create a formula by using a function which is one of Excel's many pre-written formulas that takes a value or values, performs an operation and returns a result. It is often easier to choose a pre-written formula when performing most of our statistical analysis.

OPERATORS & ORDER OF OPERATION

Many of the mathematical formulas that we use in Excel make use of the same basic arithmetic operations with which we are all familiar. Some of the arithmetic symbols we use for these **common operators** need to be entered differently and are replaced with the following:

Mathematical Operator	Symbol used in a formula
Addition	**+**
Subtraction	**–**
Multiplication	*****
Division	**/**
Exponents	**^**
Square Root	**sqrt**

Excel follows the basic **rules for order of operation** that we are all familiar with. It is important that you keep this in mind when you input a formula into Excel. You will need to be very specific in the way the information is typed. For example, $\dfrac{2+3(2)^5}{5}$ would be entered as follows (2 + 3 * 2 ^ 5)/5 in Excel. The parentheses around the expression (2 + 3 * 2 ^ 5) are important because they indicate that (2 + 3 * 2 ^ 5) is the numerator and that this entire expression is to be divided by 5.

ENTERING A FORMULA

The **following exercise** will give you some practice in **entering a formula** into an Excel worksheet.

Suppose that you want to use Excel to determine the average of your test scores for a particular course.

1) In cell A1 type the heading SCORES

2) In cells A2 through A7 enter the following test grades: 89, 73, 95, 85, 69 and 100.

3) In cell C1 type the word AVERAGE

4) **Select the cell** to the right of C2 **to enter the formula** for finding the average of these scores.

5) **Type an = (equal sign) followed by the formula**. It is often easier to use the cell address within any formula that you write rather than the actual number although you can use either one. Using a cell address is advantageous when copying a formula to other cells.

6) To find the average for our test scores we need to add all of the scores together and then divide by the number of tests taken. To accomplish this Excel we will **type** =(a2+a3+a4+a5+a6+a7)/6 in D1. Notice that that the formula used to find the average in D1 is displayed in the formula bar.

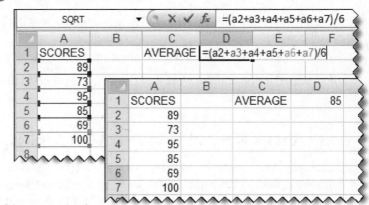

7) After typing in the formula **press Enter**.

8) You may need to use the **decrease decimal** icon to round your result off to the nearest whole number. To do this **click on the decrease decimal icon** found in the **Number group** in the **Home** tab.

Note:
Typing in the cell address for each test score in order to find a course average works well if there is a small number of cell addresses to be entered.
If we had 50 test scores in column A that we would not want to type = (A1 + A2 + A3 + …. + A50)/50. To save time we can enter the same information by typing in the cell address for a range of cells. In this case we would type in = Sum(A1:A50)/50.

COPYING A FORMULA

Suppose your instructor uses Excel to record and determine the final course averages and grades for all of her students. Rather than retype the formula used for finding the average grade for each individual student the instructor can simply copy the formula she created to find the course average for the first student into other cells. On the next page we have outlined two methods that can be used to copy a formula from one cell to another.

Method 1: (to be used when you are copying the same formula to a series of cells)

1) **Left click on the cell** that already contains the formula you want to copy.

2) **Place your cursor** on the lower right hand corner of the active cell. When your cursor changes to a cross (commonly referred to as the **fill handle**), click and hold the left mouse button down and drag the fill handle down until you have included all of the cells you wish to copy the formula to.

3) When you release the mouse button the formula will be copied and adjusted for these cells.

Fill handle

Method 2: (to be used when you are copying one piece of information to another cell)

1) **Left click** on the cell that already contains the formula you want to copy.

2) **Left click** on the copy icon found in the **Clipboard group** in the **Home** tab (or press Ctrl + C simultaneously).

3) **Activate** the cell into which you want to copy the formula.

4) **Left click** on the **Paste icon** (or press Ctrl + V simultaneously).

Try both methods on the worksheet you have created to determine which works best for you.

SECTION 1- 5: RELATIVE AND ABSOLUTE REFERENCE

Cell references are referred to as either a **relative cell reference** or an **absolute cell reference.** In each of the examples we have used so far we have used **relative cell references.** When these cell references are copied to a new location they change to reflect their new position.   This relative address feature makes it easy for us to copy a formula by entering it once in a cell and then copying its contents to other cells. An **absolute reference** does not automatically adjust when moved to another cell and is used when it is necessary to retain the value in a specific cell address when copying a formula. In an absolute reference both values in the cell address are preceded by a \$. For example, the formula = (\$B\$5 + \$C\$5 + \$D\$5 + \$E\$5) will remain unchanged regardless of the cell to which this formula is copied. The dollar sign does not signify currency but rather is used to identify that the cell is an absolute reference. A formula can contain both relative cell references and absolute cell references.

The **following exercise** will give you some practice in using both **relative cell references and absolute cell references in a formula.**

In Section 1 – 3 we entered data into an Excel worksheet outlining monthly expenses. Suppose we want to determine the percentage that each expense represents of our total expenses for the month. To do this we have to do the following:

1) Find the total of all of the expenses listed in column B. This information can be found in cell B9 as seen below.

2) Find the percentage each individual expense represents of our total expenses. To do this we divide the individual expense by the total expenses.

    a. **Enter the formula** as it is shown here   = B2/\$B\$9 in cell D2.

b. **Copy this formula** down column D. Since the top value is a relative cell reference the formula will adjust to use the cell address we are in, while the bottom value, which is an absolute cell reference, will remain constant as we would want it to.

c. Format the values in column D by using the **percentage** icon found in the **Number group** in the **Home** tab.

	A	B	C	D	E
1	MONTHLY EXPENSES			Percent of Expenses	
2	Rent	$ 950.00		60%	
3	Food	$ 250.00		16%	
4	Utilities	$ 225.00		14%	
5	Telephone	$ 55.00		3%	
6	Cable TV	$ 69.90		4%	
7	Internet	$ 39.95		3%	
8					
9	Total	$ 1,589.85			

SECTION 1-6: MODIFYING YOUR WORKBOOK

INSERTING AND DELETING ROWS/COLUMNS

Sometimes you will want to add (insert) or remove (delete) a column or row from your worksheet. If you want to create additional space in the middle of a worksheet you can insert a column or a row that will run the entire length or width of the worksheet. If you have an entire row or column that is no longer necessary you can delete the entire column or row.

To Insert a Row (or Column)

To insert a new a row or column to your worksheet

1) **Select the entire row** (or column) by right clicking on the number at the start of the row (or letter at the head of the column).

2) Left click on **Insert** found in the **Cells group** in the **Home** tab.

3) Left click on **Insert Sheet Rows (or Insert Sheet Columns)**.

The new row will be inserted *above* the selected row. A new column will be inserted to the *left* of the selected column.

To Delete a Row (or Column)

1) **Select the entire row** (or column) by right clicking on the number at the start of the row (or letter at the head of the column).

2) Left click on **Delete** found in the **Cells group** in the **Home** tab.

3) Left click on **Delete Sheet Rows (or Delete Sheet Columns)**.

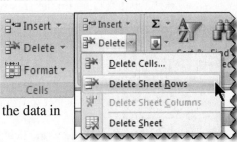

The old row (or column) is removed from the worksheet and is replaced by the data in the adjoining row (or column).

Note: Pressing the **Delete** key does not delete the selected row or column. It clears all data from the selection without moving in replacement data. Click a cell outside of the selected row (or column) to deselect it.

CHANGING COLUMN WIDTH AND ROW HEIGHT

It is not uncommon to find that we have entered something in a cell requires us to change the width of a column. Text is often cut off because the column is not wide enough to display all that we have entered into the cell. If a cell can't display an entire number or date, the cell may fill with ####### or display a value in scientific notation. (Try entering a 12 digit number into a cell.)

To Adjust the Column Width

1) **Position the cursor** on the right border of the lettered heading at the top of the column you wish to

   adjust. The cursor will change to the move pointer and will look like this

2) **Hold down the left mouse button**, dragging it back and forth to increase or decrease the column width.

3) Release the mouse button when you have a column the width you like.

There is a quick and alternative method for adjusting the column width. Begin as you did above in Step 1. **Double left click on the right border** of the column heading. This will cause the width of the column to auto-adjust.

To Adjust the Row Height

1) **Position the cursor** on the lower border of the numbered row whose size you wish to adjust. The cursor will change to the move pointer as shown above.

2) **Hold down the left mouse button**, and drag it back and forth to increase or decrease the row height.

3) Release the mouse button when you have a row at the height you like.

ADDING WORKSHEETS TO A WORKBOOK

When you open a new workbook in Excel you will see that there are always three blank worksheets available to you. This gives you the ability to do three separate problems or variations of a problem all within one workbook. While only three worksheets are presented, it is possible to have up to sixteen worksheets within one workbook. To add a worksheet to your workbook:

Method 1

To insert a new worksheet at the end of the row of existing worksheets left click on the **Insert Worksheet** tab at the bottom of the screen. A new worksheet will be added to your workbook.

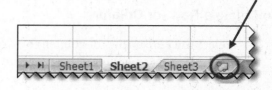

Method 2

To insert a new worksheet in front of an existing worksheet

1) Select the worksheet by left clicking on the sheet tab.

2) Left click on **Insert** found in the **Cells group** in the **Home** tab.

3) Left click on **Insert Sheet.**

The new worksheet will be inserted to the left of the existing worksheet.

RENAMING A WORKSHEET

Initially the worksheets in Excel are labeled "Sheet 1" "Sheet 2" and "Sheet 3". It is generally a good idea to rename your worksheets. This allows you to keep track of the information, data and graphs which are typically found in an Excel worksheet. To rename your worksheet:

1) **Right click on the worksheet tab** that you would like to rename

2) **Highlight Rename** and click. The sheet tab should now be highlighted.

3) Begin **typing the new name** for this worksheet.

4) Press **Enter** when you have finished.

INSERTING A COMMENT

There are many instances when it is important to annotate work done in Excel by attaching a note or comment to a cell. This is done by adding comments that can be viewed when you place the cursor over that cell.

To enter a comment into an Excel worksheet:

1) **Select the cell** that you want to add the **comment** to.

2) Left click on **New Comment** found in the **Comments group** in the **Review** tab.

3) Type your comment in the textbox that opens. When you finish typing the text, click outside the comment box.

4) An indicator (a **red triangle**) will appear in the upper right hand corner of the **cell that contains a comment**.

5) To view the comment, move the cursor over the cell that contains the note.

6) The comment will appear. This can be seen in the screen shot shown on the right.

	A	B	C	D	E	F
1	MONTHLY EXPENSES			Percent of Expenses		
2						
3	Rent	$ 950.00		60%		
4	Food	$ 250.00		16%		
5	Utilities	$ 225.00		14%		
6	Telephone	$ 55.00		3%		
7	Cable TV	$ 69.90		4%		
8	Internet	$ 39.95		3%		
9						
10	Total	$1,589.85				

E Reda
More than half of my monthly expenses goes for rent.

To Show, Edit or Remove a Comment

You will use the various icons found in the **Comments group** in the **Review** tab. Use the screen shot below to locate each of the appropriate icons.

1) To **make a comment visible** at all times activate the cell that contains the comment. Left click on **Show/Hide Comment.**

2) To **edit** a comment activate the cell that contains the comment you wish to edit. Left click on **Edit Comment**. Make your changes and then click anywhere on your worksheet.

3) To **delete** a comment activate the cell that contains the comment you wish to delete. Left click on **Delete**.

SECTION 1- 7: PRINTING YOUR EXCEL WORK

There are many occasions where you will want to print a worksheet or entire workbook that you have created in Excel.

PRINTING YOUR EXCEL SPREADSHEETS

Using Print Preview

It is *always recommended* that you use **Print Preview** before printing any work. This will give you an opportunity to make sure all of your work is presented within the printable page. It will also give you a chance to catch and correct mistakes before you print a page. To preview a worksheet before you print it:

1) **Left click** on the **Microsoft Office Button** located in the upper left hand corner of your workbook.

2) Move your cursor so that it over the **Print command** shown in the left column as seen here. Then move your cursor to the right to highlight **Print Preview.**

3) **Left click** on the **Print Preview option** which will open the **Preview group** which is shown on the next page. Once you are in this group there are several things that you can do.

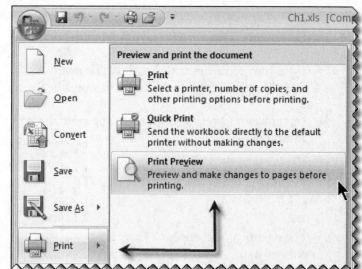

   a. To preview any other pages you might have in your Excel worksheet left click on **Next Page** or **Previous Page**. This option is only available when a worksheet contains more than one page of data or graphs.

b. To view the page margins select **Show Margins** which will display the margins in **Print Preview** view. You can adjust column width by dragging the handles at the top of the **Print Preview** page. To change margins, you simply drag the margin handles to the height and width that you want.

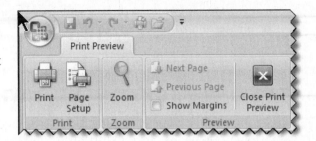

c. **Page Setup** opens a dialog box that allows us to change the orientation of the page, adjust margins and work with headers and footers. Select the tab option that you wish to work with to make changes to the page set up. Left click on **OK** when you are finished.

d. When you have completed your preview of the Excel spreadsheet you can choose to print the worksheet directly from this page by choosing **Print.**

e. To leave the **Print Preview** tab simply **left click on Close Print Preview.** This will return you back to the spreadsheet you were working with.

Setting the Print Area

There will be many occasions where you will only need to print a portion of an Excel worksheet rather than the entire worksheet.

To set the Print Area

1) On your current worksheet select the range of cells that you wish to define as your print area.

2) Left click on the **Page Layout** tab.

3) In the **Page Setup group** left click on **Print Area** and then left click on **Set Print Area.**
When you print your Excel spreadsheet, only the print area you have selected will be printed.

Note:
The print area that you set is saved when you save your workbook. If you do not want to save the print area left click anywhere on your worksheet. You can also clear the print area in the **Page Setup group.** Left click on **Print Area** and then left click on **Clear Print Area.**

To Print:

Now that we have used **Print Preview** and have set the area we wish to print we are ready to print. When we left click on the **Microsoft Office Button** and move our cursor to the **Print command** we see that in addition to Print Preview we are offered several other options.
- The **Print** option opens a new dialog box which allows you to select your printer, choose the number of copies you wish to print, and provided several other printing options. You can choose to print your **Active sheet** or the **Entire workbook** which will print all of the worksheets in your current Excel workbook

- The **Quick Print** option will sent your Excel workbook directly to a default printer as it is. You do not have the opportunity to make any choices or changes when you choose this option.

SECTION 1- 8: GETTING HELP WHILE USING EXCEL

Excel comes with a complete on-line Help feature designed to give assistance when you are having difficulty with a topic. Do not be afraid to use this feature.

Using Excel Help

Left click on the **Help** icon located in the upper right hand portion of the **Tab bar.**

- You can **Browse Excel Help.** Just select a topic heading in this section.

- You can search for help by **typing in a word or phrase** in the box and left click on **Search.** Select the topic that appears to best match your question and shows the most potential for providing you with the help you need. Left click on the topic to view **Help** information.

TO PRACTICE THESE SKILLS

It is important to practice the Excel skills introduced in this chapter before you move on. The following exercise will give you an opportunity to practice some of your new talents.

Temperature Conversions

The formula for converting degrees Fahrenheit to degrees Celsius is $C = \dfrac{5}{9}(F - 32)$. Use Excel to set up a spread sheet to do these conversions given a set of temperatures.

a) **Entering text:** Type "Degrees Fahrenheit" in Cell A1.

b) **Entering text:** Type "Degrees Celsius" in Cell B1.

c) **Entering numerical values:** In cells A2 through A11 enter the following temperatures recorded in degrees Fahrenheit:

-10°, 0°, 10°, 32°, 45°, 50°, 68°, 75°, 83°, 95°.

d) **Entering a formula:** In cell B2 enter the formula to convert from degrees Fahrenheit to degrees Celsius using a appropriate cell addresses.

e) **Copying a formula:** Copy this formula through to cell B11.

f) **Formatting data:** Format the values in column B correct to 2 decimal places.

g) **Renaming a worksheet:** Rename your worksheet "Temperature Conversion."

h) **Printing your worksheet:** Print out your worksheet.

One Final Note for Chapter 1: For the remainder of this manual we will use the instruction **"click" when we mean to left click** and will specify a **right click** when students need to **use the right mouse button**. This is consistent with terminology used in most manuals and in the Microsoft Office Help feature. .

CHAPTER 2: SUMMARIZING AND GRAPHING DATA

SECTION 2-1: OVERVIEW & ADD-INS

In this chapter, we will use the capabilities of Excel to help us summarize and graph sets of data. The sections that follow take you through step by step directions on how to create frequency distributions and histograms, as well as other types of visual models.

There is an **Add-In** that is necessary when working directly with Excel to access certain analytical features called **Analysis ToolPak**.   This Add-In can be directly accessed in Excel.

A second **Add-In** is available on the CD that comes with your text book or at the website www.aw-bc.com/triola. and is called **DDXL**. This Add-In makes it possible to do some work within Excel that may be easier to accomplish than just using Excel alone, and includes some features that are not available directly through the Excel program. We suggest you install both these Add-Ins right from the start, so they are available to you when you want to access them.

ADD-IN: LOADING ANALYSIS TOOLPAK

Before you begin your work in this chapter, make sure that you have the **Analysis ToolPak** loaded on your machine.

To see if the ToolPak has already been loaded, click on the **Data** tab, and see if you can find the **Analysis group**.

If not, follow these steps:

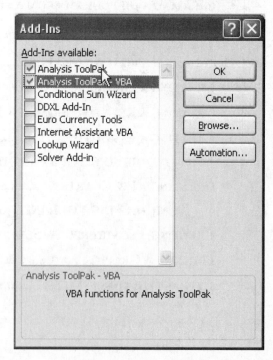

1) **Click the Microsoft Office Button** in the upper **left hand corner of your screen:**   At the bottom of the box that opens, click on **Excel Options.** In the Excel Options screen that opens, look in the sidebar on the left, and click on **Add-Ins.** This will open a **View and Manage Microsoft Add-Ins** screen. Highlight **Analysis ToolPak** and click on **GO.**

2) **Install the ToolPak:** You should see the dialog box shown. Make sure you click in the box in front of **Analysis Tool Pak** as well as the **Analysis ToolPak – VBA.** Then click **OK.** If you get prompted that the Analysis ToolPak is not currently installed on your computer, click **Yes** to install it. This will take a few minutes.

3) **Check in the Tab Bar:** After you load the ToolPak, click on the **Data** tab. You should now see the **Analysis group** as shown above.

ADD-IN: LOADING DDXL

DDXL is an add-in for Excel which you can find at the website: www.aw-bc.com/triola.   This add-in allows you to work with data which is presented in an Excel worksheet.

Follow the directions at the website to save the program to your computer, and then to set up as an add-in within Excel.

SECTION 2-2 & 2-3: FREQUENCY DISTRIBUTIONS AND HISTOGRAMS FROM DATA

In this section, we will learn how to create a frequency distribution and histogram directly from a data set. We will use the data on **Female Health** from **Data Set 1 in Appendix B: Health Exam Results**. This data set shows a number of health related measures for males and females, including pulse rate in beats per minute.

OPENING AND ORGANIZING YOUR DATA

1) **Developing Good Habits:** You should get in the habit of preserving your original data in one of the sheets in your workbook, and then working with the specific data of interest in another worksheet. This way, you can always easily retrieve it if something goes awry as you work with specific parts of the original data. You should also clearly name the worksheet tabs so that it is clear what is contained in each sheet.

2) **Load the data into Excel:** You can access the data from the CD that came with your text book. The file name is FHEALTH.XLS. You can also find the data set at the website for the textbook: www.aw-bc.com/triola.

3) **Preserve your original data in a clearly marked worksheet:** After loading the original data from your CD, or from the website, double click on the tab at the bottom of the worksheet so that the Sheet name is highlighted. Type in "Original Data."

4) **Select the data you will be working with:** For our work, we will be working with the data which shows the pulse rate of females. If you loaded the data from your CD, your data appears in column F. Select this column by clicking on the "F". This column should now be highlighted.

5) **Copy the selected data to a new worksheet:**

   a. Once you have selected the data you are going to work with, copy this data by selecting the **Copy icon** from the **Clipboard group** in the **Home** tab. You should now see your selected data in a dotted box that looks has moving edges.   (**Alternatively, you can press Ctrl + C**.)

   b. At the bottom of your worksheet next to the tab you labeled Original Data, click on the **Insert Worksheet** tab. Check to make sure your cursor is positioned at the top of column A (cell A1) and click on the **Paste icon** in the **Clipboard group** in the **Home** tab.   (**Alternatively, you can press Ctrl + V**.) Your data should be copied into your new sheet.

   c. **Rename your new worksheet:** Since we will be creating a frequency distribution and histogram for the data about pulse rates, you may want to rename your sheet with a name like "Female Pulse". Remember, to accomplish this, double click on the tab, and once it is highlighted, type in "Female Pulse".

DECIDING WHETHER TO USE DDXL OR EXCEL

You have two choices in terms of generating a histogram.

1) **Using DDXL:** If you just want to get a quick idea of how the data is distributed, and don't need the frequency distribution table, using DDXL makes sense. It is relatively easy to generate a histogram and the supporting summary statistics using this Add-In. However, you do not have

many options in terms of formatting the histogram, and have no control over the number of classes or the class width.

2) **Using Excel:** If you need both the frequency distribution and the histogram, and you want to be in control of the number of classes, as well as able to do some creative formatting of your histogram, you should use Excel. While initially using Excel can seem difficult, with practice you will find you can navigate the key steps relatively quickly.

USING DDXL TO CREATE A RUDIMENTARY HISTOGRAM WITH SUMMARY STATISTICS

1) **Select the data you want to use:** From your Excel worksheet, click on the letter at the top of the column where the data you want to use for your histogram is located, or select the cells where the data is located. You do want the title of the column included in your selection.

DDXL ▾
Summaries
Tables
Charts and Plots
Regression
ANOVA
Confidence Intervals
Hypothesis Tests
Nonparametric Tests
Process Control
Help Information
Activation Key
About DDXL
Remove DDXL

2) **Access the DDXL Add-In:** Click on the **Add-Ins** tab. As long as you have already loaded DDXL in Excel (See previous instructions for **Loading DDXL.**), you should see DDXL ▾ listed as an option in your **Add-In group**. Click on the arrow near DDXL, and select **Charts and Plots.**

3) **Select the Data in the Charts and Plots Dialog Box:** When you have clicked on Charts and Plots, a dialog box will open. Click on the drop down arrow under **Function Type**, and click on **Histogram.** Under Names and Column, you should see the word **Pulse.** Click on this, and then click on the blue arrow next to the **Quantitative Variable Box**. The word Pulse should now appear in this box. You should see a check in the box indicating that the "First Row is Variable Name". Click on **OK.**

4) **Enlarge your histogram:** You will now be in a window that says **Data Desk** at the top, and you will see a histogram and the summary statistics on the left hand side of that page. You can enlarge your histogram by first clicking on the title bar at the top of the picture to select this region, and then left clicking on the small diamond in the bottom right hand corner of the screen, holding the left click button down, and pulling out diagonally.

5) **Copying your histogram and summary statistics to another document:** Click on the top bar of either the Histogram screen, or the Summary Statistics, and then click on **Edit, Copy Window.** You can now move to the document where you want to copy the histogram, and press **Ctrl + V** to paste it. Do the same with the Summary screen. The information below was produced in DDXL.

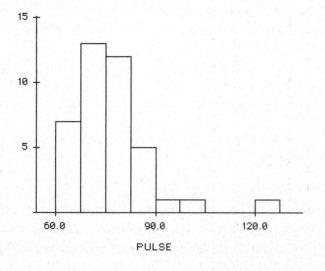

Count	40
Mean	76.3
Median	74
Std Dev	12.499
Variance	156.215
Range	64
Min	60
Max	124
IQR	12
25th%	68
75th%	80

USING EXCEL TO CREATE AN INITIAL FREQUENCY DISTRIBUTION AND HISTOGRAM

In order to create a frequency distribution, we need to indicate what **upper class limits** we want to use. Excel will refer to these upper class limits as "bins."

1) **Before using Excel:** You need to determine how many classes you want, what your class width will be, and what your lower and upper class limits will be for the particular data set. The steps are outlined in section 2-2 of your book under "Procedure for Constructing a Frequency Distribution". As outlined in the solution to the Example 1: Pulse Rates of Females in your text book, let's use 7 classes, with a class width of 10 and start with a lower class limit of 60. We would compute the **lower class limits** to be 60, 70, 80, 90, 100, 110, and 120 and the **upper class limits** to be 69, 79, 89, 99, 109, 119, and 129.   (Keep in mind that we determined these values without the help of Excel!)

2) **Set up your worksheet:** Assuming you have followed the steps in the previous section, "Opening and Organizing Your Data", you should be ready to work with the data on Female Pulse Rates in a sheet, which you have renamed "Female Pulse". Otherwise, go back and follow those steps before proceeding.

	A	B	C	D	E
1	PULSE		LCL	UCL	
2	76		60	69	
3	72		70	79	
4	88		80	89	
5	60		90	99	
6	72		100	109	
7	68		110	119	
8	80		120	129	
9	64				

3) **Set up the Class Limit column(s):** In a blank column to the left of the data, type the name "Lower Class Limit" or "LCL" in the first cell of your selected column. In the column directly to the left of that, type in "Upper Class Limit" or "UCL".   Type in the values that are listed above. (**Note:** While it is not necessary to enter the LCL into Excel, it helps us later on as we interpret our frequency table and histogram and create our Class Marks.)

4) **Access Data Analysis:** Click on the **Data Analysis icon** in the **Analysis group** in the **Data** tab. (If this feature is not available, go back to the instructions in Section 2-1 to learn how to add this feature to your machine.)

5) **Select Histogram:** In the Data Analysis Dialog Box, double click on **Histogram** (or select **Histogram** and click on **OK**).

6) **Work with the Histogram Dialog Box:** Follow the steps below to create the dialog box shown.

    a. **Collapsing the Dialog Box:** If you are going to fill in the ranges by selecting the appropriate cells in your worksheet, you will want to collapse the dialog box by first clicking on the **collapse icon** on the right hand side of

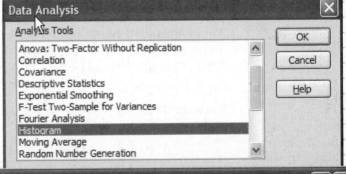

the input boxes. This will collapse the dialog box, and allow you to move freely around your worksheet. Once you have selected your input, click on the **expand icon** to go back to the dialog box.

b. **Input Range:** You need to tell Excel what data you want to sort. This is referred to as the Input Range. Since we want the frequency distribution of Pulse rates, for the "Input Range," we need to tell Excel where this data can be found. You can either type the cell range in the box with a colon separating the first and last cell (for example, type **A2:A41** if your data can be found in these cells) **OR** select the cells where this data is located in your worksheet. (Refer back to Chapter 1 if you need to review how to select a range of cells.) If you are going to select the cells, you should first click on the **collapse icon** so that you can move freely around your worksheet. Once you have selected your input, click on the **expand icon** to go back to the dialog box.

c. **Bin Range:** Excel refers to the **upper class limits** as **bins.** In the **Bin Range** box, either type in the range of cells where the **upper class limit** values are located with a colon in between cell names, or select these cells in your worksheet.

d. **Output Options:** You can either have your results positioned in your current worksheet, or in a new worksheet.

 i. To position your **output in the same worksheet**, click in the circle in front of **Output Range, then click inside the entry box**, and either type the cell location where you want your data to begin, or use the **collapse icon** and select an appropriate cell. Notice in the dialog box above, Excel will position the output for the frequency distribution starting in cell F9. **Note:** If you are going to overwrite data that already exists in your worksheet, a warning dialog box will appear.

 ii. **New Worksheet:** If you want to have your output returned in a separate worksheet, click in front of **New Worksheet Ply**, and type in a meaningful name such as "Female Pulse Histogram".

e. **Always Include Chart Output:** Since it is rare that you would want to create a frequency distribution without the connected picture, get in the habit of always clicking in front of "Chart Output". This will ensure that you create both the frequency table as well as an initial histogram.

f. **Click on OK.** You should now see a frequency distribution for your data as well as a rudimentary histogram.

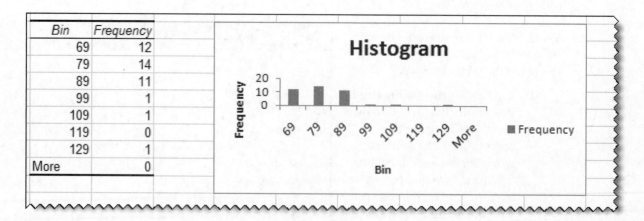

Bin	Frequency
69	12
79	14
89	11
99	1
109	1
119	0
129	1
More	0

Note:
You should remember to interpret the frequencies as the number of values that are less than or equal to the upper class limits that you typed into your bins, but greater than the previous upper class limit.
Looking at the frequency distribution for the Distribution of Pulse Rates for Females, you should interpret that there are 14 ratings between 70 and 79 (less than or equal to the bin value 79, but greater than the previous bin value 69).

Modifying the initial frequency distribution

1) **Delete the row labeled "More":** Notice that when Excel created the frequency distribution, there is an extra row labeled "More." This helps ensure that you count all the data in the set, even if you mistakenly did not include a high enough upper class limit in your bins. If you included enough upper class limits, this row should show a "0". To remove this extra row, highlight only the two cells in the frequency distribution that you wish to remove. Click on **Delete** in the **Cells group** in the **Home** tab. Notice that when you delete this row, the extra column in the histogram picture also disappeared.

2) **Rename the first column of the frequency distribution:** We want the frequency distribution to clearly indicate what the data represents, so we want to create a name other than "Bin".

    a. Click on the cell where you see the name "Bin" at the top of the column, and type in the new name you would like to use. In this example, it makes more sense to type in what the values represent: "Pulse Rates for Females".

3) **Wrap the Text:** This name would extend over a larger region than the original column width. To rectify this, after you type in the heading, click on another cell, and then re-click on the cell containing your heading. With this cell selected, click on the **Wrap Text icon** in the **Alignment group** of the **Home** tab.

4) **Change the "bin" values to Class Marks:** From your textbook, you learned that a histogram should show either the class boundaries or the class marks along the horizontal axis. If you use class boundaries, they should appear under the vertical lines of the bars. If you use class marks, they should be centered under each bar. Remember that in order to have Excel create the frequency table, you needed to tell it to use the upper class limits as the 'bins". Now that you have created the initial frequency table and histogram, you need to change the bin values in the frequency table to the class marks. (In Excel, it is too complicated a procedure to use the class boundaries in the histogram!)

5) **Using Excel to compute the class marks:** If you have entered both your lower class limits and upper class limits into Excel, it is relatively easy to use a formula in Excel to automatically compute the class marks. While you could certainly do this by hand, it's not a bad idea to get used to creating and copying formulas in Excel! You can refer back to Chapter 1 to review how to enter formulas.

    a. **In order to use a formula, you will reference cell addresses.** If your lower class limits started in cell C2, and your upper class limits started in cell D2, you could type the following formula in the first cell of your frequency table: =(C2+D2)/2 The equal sign is what tells Excel you are entering a formula. The parentheses are essential in order to perform the addition before the subtraction. When you press Enter you should see the first class mark of 64.5 in the cell where you entered the formula.

Pulse Rates for Females	Frequency
=(c2+d2)/2	12
79	14
89	11
99	1
109	1
119	0
129	1

b. **Now you want to copy the formula using the fill handle.** Click on the class mark value of 64.5. Move the cursor to the black corner of the box containing this first class mark until it turns into a "+" sign. Now hold down the left mouse key and "pull" the handle down so that you have the 6 additional boxes containing the bin (upper class limit) values "selected". When you let your mouse up, you should see that your cells have been "filled" with the updated formulas which create the class marks for each of the classes using successive lower and upper class limits. **Your histogram changes as you change the frequency table:** Notice that when you changed the values in your frequency table, the values also changed on your histogram.

Modifying the initial Histogram

You will need to modify the histogram some to make it look the way you want it to. In particular, you will want to close the gap between the bars, re-name the horizontal axis, and give the graph an appropriate title. In addition, you will want to resize your graph. Instructions to complete the revisions follow.

1) **Choose a design which closes the gap between bars:** A histogram has bars that are touching. Therefore we need to adjust the initial picture.

   a. Click anywhere inside the chart area of your original histogram. When you do this, you should see a new element called **Chart Tools** added to the Tab Bar.

   b. Click on the **Design** tab, and then on the down arrow for the **Chart Layouts group**. In the expanded box showing all the possible layouts, click on Layout 8. This layout will automatically close the gap between the bars. (You could also accomplish this by right clicking on one of the histogram bars. Then Click on **Format Data Series** in the shortcut menu that is displayed. In the dialog box that opens, move the slider in the Gap Width area to No Gap.)

2) **Create clearer borders for your bars:** Click on any of the bars. Again, notice that you see a new element called **Chart Tools** added to the Tab Bar. Click on the **Format** tab in the **Chart Tools** tab, and then in the **Shape Styles group**, click on **Shape Outline.** In the box that opens, click on **Weight**, and select the weight of the line you want to use as the border for the bars.

3) **Change the Graph Title and Horizontal Labels:** The graph title that comes up on your histogram by default is "Histogram," and the x axis label is "Bins." You will want to change the names to be more indicative of what the graph represents. Select the title or the axis name you want to change by clicking on it. Your title or axis name will have a "selection box" around it. To change the name, simply begin typing the new name you wish to use. Notice the text is typed up at the top of the worksheet, and the title in the box is changed once you press **Enter**.

4) **Resize a Region:** You would not usually leave your picture the size it appears initially. Click somewhere inside the histogram and then move your cursor around until you see the **Chart Area** tag appear. When you see that tag, click again. This selects the region you want to resize. Notice the dots in the four corners of the Chart Area, as well as in the middle of each side. Move your cursor

to the dots in the lower right hand corner of the Chart Area until you see a double headed diagonal arrow appear. Once you have this arrow visible, hold your left click button down, and drag diagonally to enlarge the Chart Area.

5) **Other Changes:** You can change the font size along either of the axes, as well as for your Chart and Axes Titles by **right clicking** near the numbers or letters you want to change. You will see a formatting box appear, and you can make many changes using the options available.   When you are done, your graph should clearly display the information below.

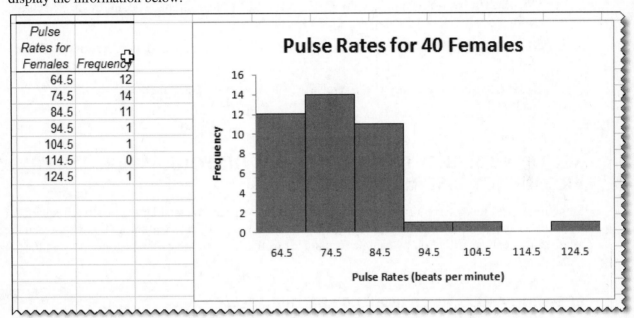

Pulse Rates for Females	Frequency
64.5	12
74.5	14
84.5	11
94.5	1
104.5	1
114.5	0
124.5	1

Pulse Rates for 40 Females

Selecting Regions in a histogram & Accessing Formatting Options

1) **Click in the white box containing the histogram, and move your mouse over different parts:** You will see "tags" come up telling you what the various regions are called. Regions include:

   - Horizontal (Category) Axis (Horizontal axis under the histogram where the "bin" values appear.)
   - Horizontal (Category) Axis Title (Initially reads Bins)
   - Chart Area
   - Vertical (Value) Axis
   - Vertical (Value) Axis Title (Currently reads Frequency)
   - Plot Area
   - Series "Frequency" Points (Gives the frequency as well as each of your class marks)
   - Chart Title (Initially reads Histogram)

2) **Click on a region and notice that a box with dots, called "handles", appear around that region.** This indicates that you have "selected" the region.

3) **Right click while a region is selected to access the formatting menu.** You have many options open to you in terms of what type of formatting changes you can make. We encourage you to "play" with the various options to create your own individualized picture.

TO PRACTICE THESE SKILLS

To really learn how to use Excel well, you need to practice the skills covered in the previous section several times before you "own them."

Notice that there are a number of exercises in section 2-2 and 2-3 of your text book which ask you to create a frequency distribution (section 2-2) and a histogram (section 2-3) using the data sets found in the back of your book. You should practice your skills with some of the paired sets below:

1) **Nicotine in Non-filtered Cigarettes:** Exercise # 19 from section 2-2, paired with exercise # 11 from section 2-3.

2) **Home Voltage Measurements:** Exercise # 21 in section 2-2, paired with exercise # 13 from section 2-3.

3) **How Long is a ¾ in. Screw?:** Exercise # 23 from section 2-2, paired with exercise # 15 from section 2-3.

SECTION 2-3: RELATIVE FREQUENCY HISTOGRAM FROM A GIVEN FREQUENCY DISTRIBUTION

Sometimes we already have a frequency distribution, and we want to consider the relative frequency distribution and relative frequency histogram. The steps below will take you through how you can create the relative frequency distribution by using a formula, as well as how to create the histogram using the Chart Icon.

CREATING A RELATIVE FREQUENCY DISTRIBUTION

Suppose we had the frequency distribution from the previous section, and we wanted to create a relative frequency distribution. We need to add a column for relative frequencies to our frequency distribution. The relative frequency for a particular class will be equal to the number of values in that class divided by the total number of values in the data set. To set up this column in the frequency distribution follow the steps below.

	A	B
1	Pulse Rates for Females	Frequency
2	64.5	12
3	74.5	14
4	84.5	11
5	94.5	1
6	104.5	1
7	114.5	0
8	124.5	1
9		

1) **Type the table into a new worksheet, or use Paste Special to move the table to a new worksheet:** If you don't have the frequency table already available in Excel, type the table in as shown. If you have access to the table from the previous work we did, select the frequency table and then press **Ctrl + C** (Hold down the Ctrl key and then press the C key.) Insert a new worksheet by clicking on the **Insert Worksheet icon** at the bottom of your worksheet.  . Position your cursor in cell A1, and then click on the down arrow under the **Paste icon** in the **Clipboard group** in the **Home** tab. Select **Paste Special.** Using Paste Special is necessary here since the original frequency distribution had the class marks based on formulas. If you tried to paste these into a new worksheet, you would see an error "#REF!" which tells you that you have an invalid cell reference error.

2) **Create the Total:**

    a.   In a cell under the entries in the first column of your frequency distribution, type the word "Total."

    b.   Staying in the same row as where you entered "Total," click in the cell under the frequency column.

    c.   Now click on the auto sum button on in the Editing region of the Home Ribbon. You should see a screen like the one shown. Make sure the cells selected are the ones you wanted to add. If so, press **Enter.** If not, select the range of appropriate cells before pressing **Enter.** You should now see the total number of data values in your data set, which should be 40.

	A	B	C	D
1	Pulse Rates for Females	Frequency		
2	64.5	12		
3	74.5	14		
4	84.5	11		
5	94.5	1		
6	104.5	1		
7	114.5	0		
8	124.5	1		
9				
10	Total	=SUM(B2:B9)		
11		SUM(number1, [number2], ...)		
12				

3) **Create the Relative Frequency Column:**

    a.   At the top of the next column of your frequency distribution, type in the name "Relative Frequency", and press **Enter.**

    b.   Re-click on the heading. Click on **Format** in the **Cells group** in the **Home** tab and select **Format Cells.** Then under **Alignment,** make sure there is a check mark in front of "Wrap Text." Then click on **OK**.

    c.   Since we typically want to see the relative frequencies displayed as percents, change the formatting of the column.

        i.   Click on the letter at the top of the column to select the column, or select only the cells where your values will appear.

       ii.   Click on the % symbol in the **Number group** in the **Home** tab.

	A	B	C
1	Pulse Rates for Females	Frequency	Relative Frequency
2	64.5	12	=b2/b10
3	74.5	14	
4	84.5	11	
5	94.5	1	
6	104.5	1	
7	114.5	0	
8	124.5	1	
9			
10	Total	40	
11			

4) **Create your formula:** In the cell immediately to the right of your data for the first class, enter the formula that will divide your frequency count for a class by the total number of data values. For example, if your first upper class limit is in cell A2, your frequency count for that class is in B2 and your total frequency count is in B10, you would type: =b2/b10, and press **Enter.** Notice that b2 (where your first frequency count can be found) is a relative address, while b10 (where the total of all frequencies can be found) is an absolute address, indicated by the $ signs. This will allow the numerator to be updated when you copy the formula, but will keep the total number of values (in this case, 40) fixed.

5) **Copy this formula:** Use the fill handle to fill in the remaining cells in your table. You should see a table similar to the one shown.

	A	B	C
1	Pulse Rates for Females	Frequency	Relative Frequency
2	64.5	12	30%
3	74.5	14	35%
4	84.5	11	28%
5	94.5	1	3%
6	104.5	1	3%
7	114.5	0	0%
8	124.5	1	3%

CREATING A HISTOGRAM FROM A FREQUENCY TABLE USING THE COLUMN CHART ICON

Now that we have the relative frequency table, we want to create the relative frequency histogram.

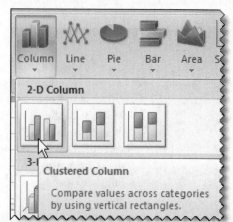

1) **Access the Column Chart Icon:** To access the **Column Chart icon**, you can click on the **Insert** tab, and then click on **Column** in the **Charts group**. Select the first type of Column Chart in the menu. A blank Chart Area will appear in your worksheet.

2) **Select the Data you want to graph:** Right click somewhere inside the Chart Area, and choose **Select Data**. The **Select Data Source** box will open.

   a. **Select the values that represent the vertical axis:**

   Under the **Legend Entries (Series)** box, click **Add.** The **Edit Series** dialog box will appear. Use **the collapse and expand icons** to help in selecting the appropriate cells for this dialog box. Select the cell where you have the word Relative Frequency for the Series Name, and select the cells where the percentages are for the Series Values. Then click on **OK.**

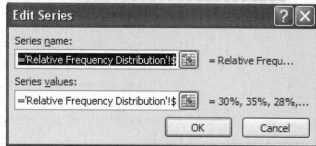

   b. **Select the values that represent the horizontal axis:** Under the **Horizontal (Category) Axis Labels** box, click **Edit.** Using the **collapse icon**, select the cells where your class marks are listed, and after clicking on the **expand icon**, click on **OK.** This takes you back to the **Select Data Source** box, where you should again click on **OK.** You should see the chart below.

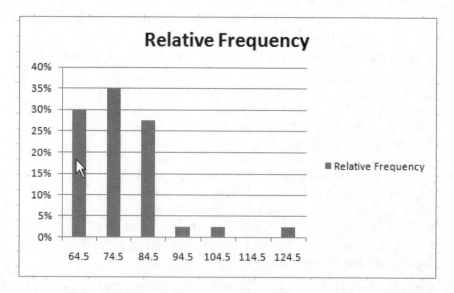

3) **Modify your Chart:** Refer back to **Modifying the Initial Histogram** directions in sections 2-2 & 2-3 of this manual to recall how to:

    a.   Chose a Design that eliminates the Gap between the bars

    b.   Create clearer borders for your bars.

    c.   Change the Graph Title and Horizontal Labels

    d.   Resize a Region

    e.   Other Changes

Your final relative frequency histogram should look similar to the one shown below.

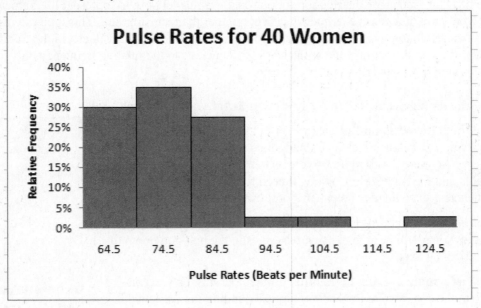

TO PRACTICE THESE SKILLS

To practice making relative frequency tables and relative frequency histograms, you can work with the frequency distributions you made in section 2-2 of this manual. For each of these frequency histograms, follow the directions in this section to create the relative frequency histogram, and then use the **Column Chart icon** to create the corresponding relative frequency histogram.

SECTION 2-4: STATISTICAL GRAPHS

We can use Excel to create a number of other types of graphs, including Frequency Polygons, Bar Graphs, Pareto Charts, Pie Charts and Scatter Diagrams. The instructions contained in this section will help you create these graphs.

CREATING A FREQUENCY POLYGON

Once we have the frequency distribution for our data, we may choose to represent it graphically as a frequency polygon or a relative frequency polygon rather than a histogram. This type of graph uses line segments connected to points located directly above your class midpoint values. In the frequency polygon, the heights of the points correspond to the class frequencies. In the relative frequency polygon, the relative frequencies are represented on the vertical scale.

Let's use the data given in Table 2-2 in section 2-2 of your text book.

1) **Type the information into Excel:** You can type this information directly into Excel, using the class midpoints rather than the listed classes. (See step 3 to find how to create the class midpoints using a formula and the fill handle.) You will need to manually enter the frequencies, since there is no pattern for those values.

	A	B
1	Pulse Rate	Frequency
2		
3	64.5	12
4	74.5	14
5	84.5	11
6	94.5	1
7	104.5	1
8	114.5	0
9	124.5	1
10		

2) **Include an extra row at the beginning of the data:** By inserting an extra row, you will be able to extend the lines in the polygon down to the horizontal axis.

3) **Using a formula and the fill handle:** Remember, just as you can move to the consecutive lower class limits by adding the class width, you can do the same thing with the class midpoints. You can use a formula and the fill handle to generate the list of class midpoints once you have the first one entered.

   a. **Enter the initial class midpoint:** Remember, the class midpoint for a particular class can be computed as (Lower Class Limit + Upper Class Limit)/ 2. For this particular data, the first class midpoint would be $(60 + 69)/2 = 64.5$. Type this into cell A3 to create the table shown.

   b. **Create a formula:** In the Excel page shown, click in the cell where you want to enter your formula. (In this example, click in cell A4.) Type the formula shown. The A in the cell address does not have to be capitalized. Then press **Enter**.

	A	
1	Pulse Rate	F
2		
3	64.5	
4	=A3+10	

   c. **Use the fill handle to create the remaining class midpoints:** Click back in the cell where you have just entered the formula, and use the fill handle to create the remaining class midpoints.

	A	
1	Pulse Rate	F
2		
3	64.5	
4	74.5	
5		
6		
7		
8		
9		
10		

4) **Select your Data and Access the Line Graph Option:**

   a. **Select the entire table that you created**: When you select your data, include an additional row below the values you entered. This will allow you to extend the lines to the horizontal axis. If you entered your data as shown above, you would select cells A1 through B10.

   b. **Access the Line Graph Option:** Click on the **Insert** tab, and then click on the **Line icon** in the **Charts group**. You will see options for 2-D Line graphs. Click on the first option in the second row for this particular exercise, since we want to see a series of points connected with straight lines. You will see a chart appear in your Excel worksheet. Don't worry that it doesn't look like what you might expect! We will make appropriate adjustments in the next steps.

   c. **Access the Select Data Source Dialog Box:** Right click anywhere within the chart region and click on the **Select Data** option. This opens up the Select Data Source Dialog Box.

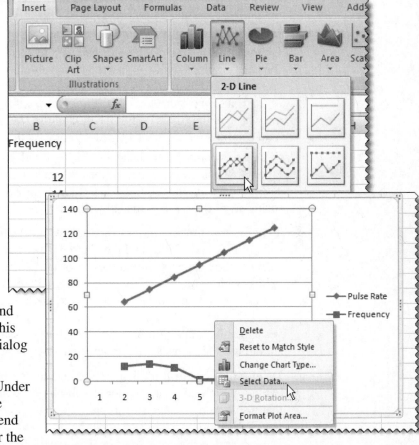

   d. **Revise Data in the Dialog Box:** Under the **Legend Entries**, click on Pulse Rate, and click on **Remove**. (Legend Entries refer to the ouput values, or the values you want along the vertical axis, which in this case are your frequencies. Then click on **Edit** under the **Horizontal (Category) Axis Labels.** This is where you will tell Excel what values to use along the horizontal axis. You will see the Axis Labels Dialog box open. Click on the **collapse icon** ▣ on the right hand side of the input box. This will collapse this dialog box, allowing you to select data from your Excel worksheet. Select the data under pulse rates, including the blank cells above and below the class midpoints. If you set your data up as shown on the previous page, this should be in cells A2 through A10. When you

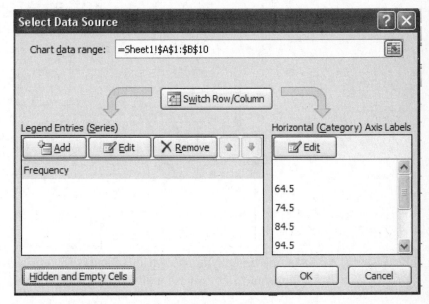

have selected your data, you should see =Sheet1!$A$2:$A$10 in the box. Then click on the

expand icon 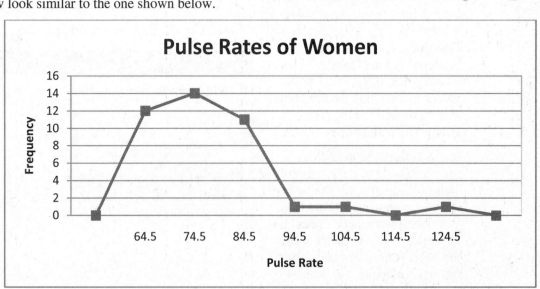 on the right hand side of the input box to expand the box. Click on **OK.** Finally, click on the **Hidden and Empty Cells** button in the lower left hand side of the dialog box, and click in front of the Zero option. This step is necessary in order for your lines to begin and end along the horizontal axis. Then click on **OK.** Your dialog box should now look like the one shown. Click on **OK.**

e. **Choose the Design:** You now need to select a design for your graph. When you insert a chart in your worksheet, a **Chart Tools** tab should appear at the top of the Excel Worksheet. If this is not there, simply click anywhere in the Chart Region. Click on the **Design** tab. You can now see a **Chart Layouts group**. Click on the down arrow on the right hand side to see all the possible layouts. Click on the option in the bottom left hand corner to select the design that allows you to label your axes, as well as include a title for you graph (Layout 10).

f. **Type in appropriate names for the axes and the title:** Click on each of the "Axis Title" and the Chart Title prompts so that a selection box appears around these words. (See the picture below, which shows a selection box around the Frequency legend.) Click inside the selection box, and delete the title that appears. Type in the appropriate name for the horizontal and vertical values, and after you finish typing, press Enter. You should use the labels as shown in Figure 2-5 of your text book.

g. **Delete the Legend Box:** Unless the legend box really helps explain your picture, it is better to "un-clutter" your graph. Click on the legend so that a selection box appears around it, and then press delete. Your graph should now look similar to the one shown below.

Pulse Rates of Women

5) **Spend some time trying the various formatting options:** Excel offers you may different options for formatting your charts. If you right click on any of the numbers along your horizontal or vertical axis,

you have the opportunity to format the axis differently. If you right click within the plot area, you will see the options available there. Likewise, you can right click within the graph region itself. Spend some time becoming familiar with the different regions of the graph, and the formatting options available to you.

6) **Resizing your chart:** In Excel, notice the dots that appear in the four corners of your chart, as well as in the middle of each side. Move your cursor near any of these dots until it changes into a double headed arrow. Hold down your left mouse button, and pull up, down or diagonally to resize the chart. You may notice that as you resize your chart, the increment between the axis values change. You can always control the values that appear on any axis that is automatically set by formatting that axis.

CREATING A RELATIVE FREQUENCY POLYGON

Suppose you want to create the relative frequency for the frequency table in Table 2-2 of your text book. If you already have the frequency polygon created in Excel, you can simply make an additional column in your data, and then make the appropriate modifications to your frequency polygon graph (directions below). Otherwise, you will follow a similar process as that included in the section on Creating A Frequency Polygon, after you enter the relative frequencies instead of the frequencies in Excel.

Let's work with the Excel sheet we used to create the frequency polygon in the previous section of this manual.

1) **Create a column for relative frequencies:** If you already have the Pulse Rate and Frequency column for Table 2-2 entered into Excel, we can just add a column representing the relative frequencies.

   a. **Create the Sum of the Frequencies:** Click in a cell that is below the data entered in your frequency column. Make sure you leave at least one cell blank between the last frequency and where you click so that we can have the graph come down to the horizontal axis at the end. In the **Home** tab, look to the far right, and you should see the **Editing group**. Click on the symbol that represents summation.

   Σ ▾ Excel should automatically select the cells directly above the cell where you inserted the summation command. If so, just press **Enter**. If the cells are not selected, simply select the appropriate cells and then press Enter. The sum of the frequencies for this example should be 40.

   b. **Create a column for the relative frequencies:** Click in the cell directly to the right of the first frequency of 12. Enter the formula: =b3/b11 if your worksheet is set up as the one shown. Otherwise, use appropriate cell addresses for the (frequency) / (sum of frequencies). **Notice that the cell where the frequency is located should be a relative cell address, while the cell where the sum**

	A	B	C	D
1	Pulse Rate	Frequency		
2				
3	64.5	12		
4	74.5	14		
5	84.5	11		
6	94.5	1		
7	104.5	1		
8	114.5	0		
9	124.5	1		
10				
11		=SUM(B3:B10)		
12		SUM(number1, [number2], ...)		

	A	B	C
1	Pulse Rate	Frequency	Relative F
2			
3	64.5	12	30.0%
4	74.5	14	35.0%
5	84.5	11	27.5%
6	94.5	1	2.5%
7	104.5	1	2.5%
8	114.5	0	0.0%
9	124.5	1	2.5%
10			
11		40	

is located should be an absolute cell address, as indicated by the dollar signs. Press Enter once you have entered this formula. Depending on the formatting that is set for that cell, you may see a % or you may see a decimal value. If you see a decimal value, click back in the cell where your value appeared, and in the **Number group**, click on the %. You can use the Increase Decimal or Decrease Decimal to change how many decimal places are shown in your answer. Show your answer with 1 decimal place. Now use the Fill Handle to fill in the remaining values.   Your worksheet should look like the one shown.

2) **Update the Data in your Frequency Polygon:** Assuming you have a frequency polygon already created, right click in the Chart Area, and click on **Select Data**.

a. **Edit the data in the Legends Entry Area:** Since the only data we want to change is the data along the vertical axis (we want to change it from frequency to relative frequency), click on the **Edit** button under the **Legend Entries** box. Click on the **collapse icon** by "Series Name", and select the title of the column containing Relative Frequency. Then click on the **expand icon**. Now click on the **collapse icon** in the "Series Values" box, and select your relative frequencies. Remember to select the

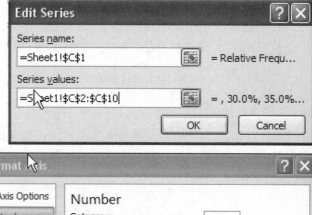

empty cells above and below the percentages. Again, this will allow your polygon to start and end on the horizontal axis. If you have your worksheet set up in the same cells as the one shown above, your dialog box should now appear as the one shown. Click on **OK** to accept the changes to the Series Data, and then click on **OK** in the Select Data Source Dialog box.

b. **Make appropriate changes to your chart:** You may need to format the numbers shown along the vertical axis so they show up as percents rather than decimals if you prefer that style. You may also have to click on the vertical axis title, and type in Relative Frequency rather than Frequency. To format the numbers on the vertical axis, right click on any of

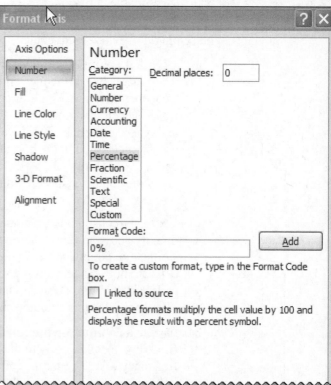

the numbers, and select **Format Axis.**   In the Format Axis Dialog Box, click on Number, and under Category, choose Percentage. Change your number of decimal places to 0. Then click on **Close.**   Your graph should now look like the one shown below.

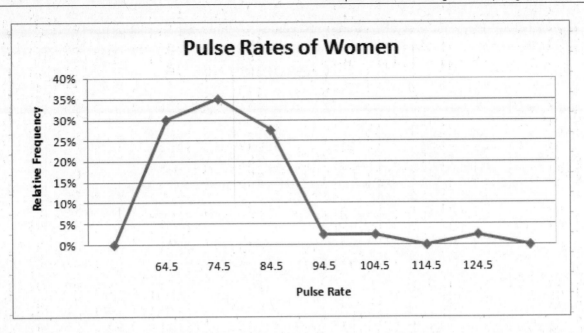

CREATING A PARETO CHART

We can use Excel to sort qualitative data in order of decreasing frequencies, and then call on the **Chart** option to create a Pareto chart.

1) **Enter the following table into a new worksheet in Excel: Make sure you use adjacent columns!** You must have the data in contiguous columns in order to properly have the information sorted. Again, get in the habit of naming your worksheets. Double click on the worksheet tab, and type Pareto to indicate that this worksheet contains your work to create a Pareto chart.

Complaints Against Phone Carriers	Frequency
Rates and Services	4473
Marketing	1007
International Calling	766
Access Charges	614
Operator Services	534
Slamming	12478
Cramming	1214

2) **Sort the Data:** We first need to sort the data in order of decreasing frequency. Click on any one cell containing a value in the Frequency column. Do **NOT** select the entire column! Click on the **Data** tab in the Tab Bar, and in the **Sort and Filter group,** click on **Sort** from largest to smallest. Once you have clicked on the **sort icon**, your table should be rearranged so that the values now go from largest to smallest, and the labels are still matched with the appropriate data value.

	A	B
1	Complaints Against Phone Carriers	Frequency
2	Slamming	12478
3	Rates and Services	4473
4	Cramming	1214
5	Marketing	1007
6	International Calling	766
7	Access Charges	614
8	Operator Services	534

3) **Select the Data and Use the Chart option to create the graph**: Select the entire table shown, including the headings. Then click on the **Insert** tab, then click on **Bar** in the **Charts group**. At the bottom of the box that opens, click on **All Chart Types.** This opens the Insert Chart Dialog Box. Click on Column, and choose the first type of graph shown. Then click on **OK.** When you are finished, your chart should look something like

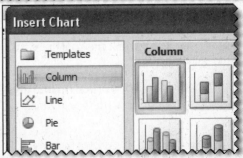

the one below. Please make sure you follow procedures to refine your original picture!

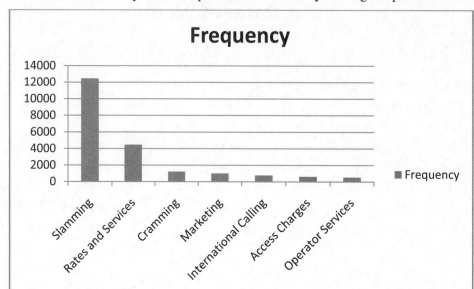

Modifying Your Pareto Chart

Once you have your basic picture, you should make further modifications to "professionalize" your chart. You should change the title to the graph, and close the gap between the columns.

1) **Change the Title:** Click on the word **Frequency** at the top of the chart. You will see a selection box around the word. Type the title: Complaints Against Phone Carriers. You will see the typing up in the input region at the top of your worksheet, and it will appear as the Chart Title when you click **Enter.**

2) **Chose the Chart Layout that closes the gap between the bars:** Click anywhere inside the Chart Area so that the **Chart Tools** tab is available on the Tab Bar. Click on the **Design** Tab.   You will see a **Chart Layouts group.** Click on the down arrow on the right hand side ▾ to see the options available. Choose Layout 8 to close the gaps between the bars. (You could also right click on a bar, choose **Format Data Series**, and move the slider in the dialog box that appears to No Gap to close the gaps.)

3) **Define your columns:** If you want to be able to more clearly see an outline of each column, click on one of the columns so that you see selection boxes around each of the columns. Then click on **Format** in the **Chart Tools** tab and in the **Shape Styles group**, click on **Shape Outline.** You can now choose the weight of the outline you want to use.

4) **Consider whether you want data labels:** Sometimes it is helpful to see how many values are actually in each column. If you want this data displayed on your chart, right click on any one of the bars, and click on **Add Data Labels.** Once you have the labels on your chart, you can format the font size, color, etc. by right clicking on one of the labels.

5) **Play with other formatting options:** The best way to become familiar with the options available in Excel is to spend some time playing with different menus. You

have a large number of ways that you can personalize your graph. Do be aware though, that oftentimes, simpler is better! Notice how the graph below is simple and easy to read.

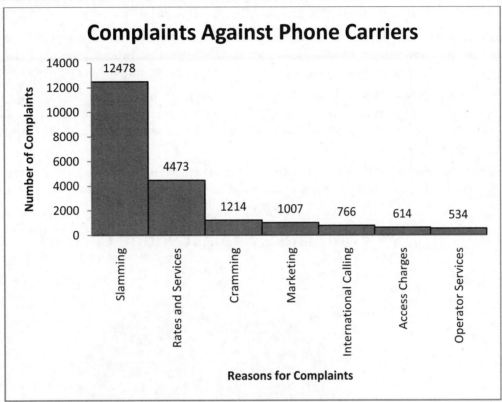

CREATING A PIE CHART

We may decide that we want to graphically show qualitative data on a pie chart, rather than as a Pareto chart. Let's work again with the information on Phone Company Complaints. The original data is listed below.

1) **Enter the Data:** Enter this information into two adjacent columns in a new worksheet, or copy it into a new worksheet if you have already entered it for the Pareto chart. **Do not leave a column in between the sources and frequencies.**

2) **Select the Data:** Select the cells containing both the column headings and the information you want to appear in the Pie Chart.

3) **Use the Pie Chart icon to create the Graph:** Click on the **Insert** tab and then click on the **Pie icon** in the **Charts group** . You will see an initial Pie Chart appear in your worksheet.

Complaints Against Phone Carriers	Frequency
Rates and Services	4473
Marketing	1007
International Calling	766
Access Charges	614
Operator Services	534
Slamming	12478
Cramming	1214

4) **Choose a Design:** Click anywhere inside the Chart Area so that the **Chart Tools** tab is available on the Tab Bar. Click on the **Design** tab of the Chart Tools tab. In the ribbon that opens, you will see the **Chart Layouts group.** Click on the down arrow on the right hand side ▼ to see the options available. The Pie Chart below uses Layout 1, with the Title changed, and the selecting the gray tone graph in the **Chart Styles group** in the **Design** tab.

Modifying Your Initial Pie Chart

You should experiment with resizing and reformatting the various parts of your chart. It is recommended that you resize the pie itself so that it takes up more of the Chart Area.

1) **Resize the circle:** Click inside the Chart Area, and move your cursor close to the circle until you see the tag **Plot Area** appear. Click once you see this tag appear, and you will see a rectangle with handles surround your circle. You can now resize the circle by pulling on one of the handles.

2) **Format the Data Labels:** Move your cursor over the labels until you see the tag **Data Labels**. Click once you see this tag appear. You will see handles appear by the various labels. Right click, and in the menu box which appears, highlight and click on **Format Data Labels**. Experiment with the different options available.

3) **Format the Chart Title:** Move your cursor over the legend until you see the tag **Chart Title**. Click once you see this tag appear. Right click, and then click on **Format Chart Title**. Experiment with the different options.

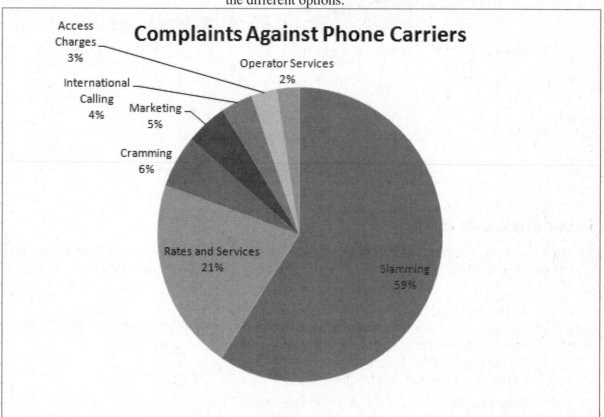

CREATING A SCATTERPLOT (SEE CHAPTER 10)

If you want to plot paired data (x, y) with a horizontal x – axis and a vertical y –axis, see the instructions in Chapter 10 of this manual.

PRINTING A GRAPH

You can elect to print only the final graph that you create.

1) **Select the Graph:** Select the graph you wish to print by positioning your cursor somewhere in the **Chart Area**, and clicking once.

2) **Preview the Graph:** Click the Microsoft Office Button [icon] in the upper left hand corner of your screen.   Place your cursor on the **Print icon**, and then move your cursor over to click on the **Print Preview** option. You should now see a screen shot of what your graph will look like when you print it. You will notice that a **Print Preview** tab appears at the top of the page.

   a. You can change the page margins by first clicking on the box in front of **Show Margins** in the **Preview group** of the **Print Preview** tab.   Position your cursor on the dotted page guide you want to move. Your cursor should change into a double-headed arrow bisected by a straight line segment. Hold your left click button down and drag your cursor to the position you want the margin to be. Then release the mouse. You should see your page margins change in the direction you moved.

   b. If you want to modify your graph, close the **Print Preview** window and make the changes you want to the graph.

   c. Re-check your picture in the **Print Preview** window again by following the instructions given in 2) above. When your picture is as desired, click on the **Print icon** found in the **Print group.**

TO PRACTICE THESE SKILLS

1) **Frequency Polygon:** You can work on Exercise # 12 from section 2-4.

2) **Pareto Charts:** You can work on Exercises # 13 and # 16 from section 2-4

3) **Pie Charts:** You can work on Exercises # 14 and # 15

CHAPTER 3: DESCRIBING, EXPLORING, AND COMPARING DATA

CHAPTER 3: DESCRIBING, EXPLORING, AND COMPARING DATA

SECTION 3-1: OVERVIEW

In this chapter, we will learn how to use Excel to create the basic statistics that describe important numeric characteristics of a set of data. Since technology allows us to easily create important values without having to memorize formulas or perform complex arithmetic calculations, we can spend more time making sure we know how to interpret and use these values to understand our data.

SECTION 3-2 & 3-3 MEASURES OF CENTER AND VARIATION

OPENING AND ORGANIZING YOUR DATA

Developing Good Habits: Again, you should get in the habit of preserving your original data in one of the sheets in your workbook, and then working with the specific data of interest in another worksheet. This way, you can always easily retrieve it if something goes awry as you work with specific parts of the original data. You should also clearly name the worksheet tabs so that it is clear what is contained in each sheet.

We will work with the data from the Chapter Problem: **Data Set 8 in Appendix B** (Word Counts by Males and Females). Follow the key steps below. The details for these steps are given in Chapter 2, section 2-2.

1) **Load the data into Excel.**

2) **Preserve your original data in a clearly marked worksheet.**

3) **Organize the data you will be working with.** The data set contains word counts from each of 6 different sample groups. We want to combine the sample data into two groups: one for data on females and one for data on males.

    a. **Copy the data into Sheet 2:** Copy and paste the data into Sheet 2, and then rename Sheet 2 appropriately. You could call it Combined Columns, since that is what we will create.

    b. **Change the column titles:** Your original column titles are 1M, 1F, etc. Rename 1M as Male and 1F as Female.

    c. **Combine the data:** Use drop and drag (See instructions in Chapter 1), to create one column with all the male data and one column with all the female data. You do not want to include the column labels, and once you are done moving the data in each column, you can delete the column headings 2M, 2F, etc.

PRODUCING A SUMMARY TABLE OF STATISTICS USING DESCRIPTIVE STATISTICS TOOL

1) **Access the Data Analysis feature:** Click on the **Data** tab, and then select the **Data Analysis option** in the **Analysis Group. Note:** If Data Analysis does not show up as an option, you need to load this as an Add-In in Excel. Follow the directions in section 2-1 of this manual.

2) **Select Descriptive Statistics**: In the Data Analysis Dialog Box, double click on Descriptive Statistics (or select Descriptive Statistics and click on **OK**).

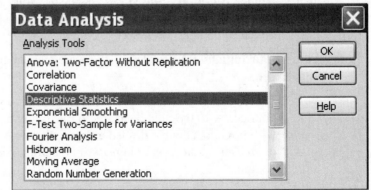

3) **Work with the Descriptive Statistics Dialog Box:** Suppose we want the statistics on the data for males.

a. **Input Range:** You need to tell Excel what data you want to use. Suppose we want to find the descriptive statistics for Word Count of Males. We need to indicate what data we want to use. There are a number of options for choosing the data. You can: 1) Type in the beginning and ending cells where your data is located, separating the 2 cells with a colon, 2) Select these cells in your worksheet. To select the cells, first click on the **collapse icon** on the right hand side of the input box. Once you have selected your data, click on the **expand icon**. 3) You could also just click on the column letter to select the entire column. If you do this, you

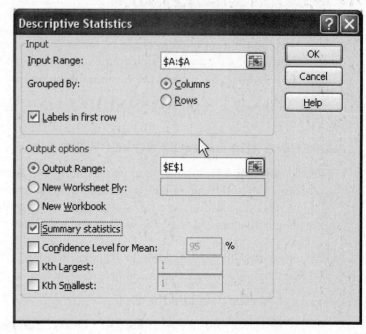

MUST indicate that the first row contains a Label, so click in front of the box to indicate this. **In the worksheet used to create this manual, this data could be found in column A, and we clicked on the letter A to select the entire column**

b. **Output Range:** You can either have your results positioned in your current worksheet, or in a new worksheet.

   i. **To position your** output in the same worksheet, click in the circle in front of Output Range, and then click inside the entry box, and either type the cell location where you want your data to begin, or use the **collapse icon** and select an appropriate cell. Notice in the worksheet used to create this manual, the statistical output would begin in cell E1.

   ii. **New Worksheet:** If you want your output returned in a separate worksheet, click in front of New Worksheet Ply, and type in a meaningful name such as "Male Word Count".

c. **Choose Summary Statistics:** Make sure you choose the Summary Statistics option, and then click on OK. You will see your Summary Statistics displayed starting in the cell you indicated on your worksheet.

4) **Adjust your initial output:** Clearly, there are adjustments that need to be made to the initial output from Excel.

a. **Resize your Columns:** You want to be able to see the full words in the first column. Move your cursor to the right hand side of the column name where you want to resize. Your cursor will turn into a double headed arrow with a straight line segment through it. When it does, double click, and your column will be automatically sized to fit the longest word in the column.

b. **Change the Title and Wrap the Text:** Click on the title Male (or Column 1 if you only selected the data and not the column label) and type in "Statistics for Word Count of Males." Press Enter and then click back on your title. In the **Home** tab, select the **Wrap Text icon** from the **Alignment group.** If you want to also center the title over both columns, select the 2 cells at the top of the table, and then click on the **Merge and Center icon** (found directly below the **Wrap Text icon**) in the **Alignment group.**

c. **Delete measures that you don't need:** You may want to delete some of the standard choices from your list. Select the two cells that contain the information on Kurtosis, and press **Delete**. Do the same for Standard Error.

d. **Sort the data:** You now have empty rows in your table. Select all the information in the two columns of your table. Do not include the title in your selection box. Then click on the **Sort icon** in the **Editing group** in the **Home** tab. You can then choose in alphabetical order, or in reverse alphabetical order. You will now see the information presented in the order you chose with the extra rows eliminated.

e. **Remove the bottom bar:** To remove the bottom bar, select the two cells containing the bottom border. Click on the down arrow by the **Border icon** in the **Font group** of the **Home** tab, and select the **No Border icon** in the displayed table. Your statistics should appear as shown in the table.

Statistics for Word Count of Males	
Count	186
Maximum	47015.63
Mean	15668.53
Median	14290.14
Minimum	694.7027
Mode	#N/A
Range	46320.92
Sample Variance	74520827
Skewness	1.0451
Standard Deviation	8632.545
Sum	2914347

Interpreting the Output in the Descriptive Output Table

Below is a brief description of each of the measures included in Descriptive Statistics.

- **Mean:** The arithmetic average of the numbers in your data set.

- **Standard Error:** This is computed by using the formula $S / \sqrt{n}$ where S is the sample standard deviation and n is the number of observations.

- **Median:** This is the data value that splits the distribution in half. To determine the value of the median, the observations are first arranged in either ascending or descending order. If the number of observations is even, the median is found by taking the arithmetic average of the two middle values. If the number of observations is odd, then the median is the middle observation.

- **Mode:** This is the observation value associated with the highest frequency. **Caution:** Three situations are possible regarding the mode: 1) if all values occur only once in a distribution, Excel will return #N/A. 2) If a variable has only one mode, Excel will return that value. 3) If a variable has more than one mode, Excel will still return only one value. The value used will be the one associated with the modal value that occurs first in the data set. To check the accuracy of the mode, it would be wise to create a frequency distribution.

- **Standard Deviation:** This is computed using the formula: $S = \sqrt{\dfrac{\sum (X - \overline{X})^2}{n - 1}}$

- **Sample Variance:** This is the standard deviation squared.

- **Kurtosis:** This number describes a distribution with respect to its flatness or peakedness as compared to a normal distribution. A negative value characterizes a relatively flat distribution. A positive value characterizes a relatively peaked distribution.

- **Skewness:** This number characterizes the asymmetry of a distribution. Negative skew indicates that the longer tail extends in the direction of low values in the distribution. Positive skew indicates that the longer tail extends in the direction of the high values.

- **Range:** The minimum value is subtracted from the maximum value.

- **Minimum:** The lowest value occurring in the data set.

- **Maximum:** The highest value occurring in the data set.

- **Sum:** The sum of the values in the data set.

- **Count:** The number of values in the data set.

CREATING PARTICULAR SAMPLE STATISTICS USING THE INSERT FUNCTION ICON

If you just want to know particular values, without producing the entire table of Descriptive Statistics, you can use the **Insert Function icon** in the **Function Library group** of the **Formulas** tab, and select just the options that you want to use.

1) **Decide which values you want to compute:**
   Corresponding to ideas presented in section 3-2, we will compute the Mean, Median and Mode for **Data Set 8 in Appendix B**. We will find the values for each of the groups contained in that data set: Males and Females

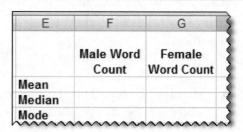

E	F	G
	Male Word Count	**Female Word Count**
Mean		
Median		
Mode		

2) **Create 2 columns for the data:** Refer back to the instructions under *Opening and Organizing Your Data* at the beginning of section 3-2 & 3-3 to see how to combine the various columns of data in the original data set into just 2 columns.

3) **Create column and row headings:** Type in the information shown to the right. Use the **Wrap Text icon** in the **Alignment group** of the **Home** tab to create the look in the table shown.

4) **Create the first Mean for Males:**

   a. Select the cell next to **Mean**, and directly under the title **Male**. Click on the **Formulas** tab, and then on the **Insert Function icon** in the **Function Library group.** In the dialog box that appears, click on the down arrow by the category box, and select Statistical. In

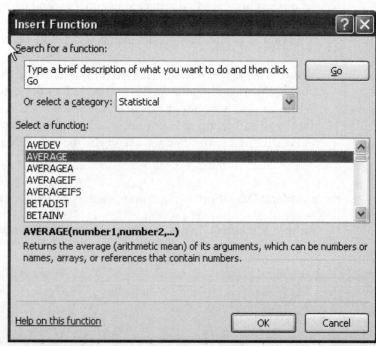

the function list that appears, click on **AVERAGE** and then click on **OK.**

    b. **Fill in the Function Arguments Box:** On the line with the input box by **Number 1**, click on the **collapse icon**, and select the data in your worksheet for which you wish to compute the mean. In the worksheet used for this manual, that data was in cells A2:A187. Then click on the **expand icon** to re-expand your dialog box. Click on **OK.** You should see the average for your data returned in the cell you had your cursor positioned in on your worksheet. **Alternatively:** As long as you do not have anything in the column below where your word counts for males is located, you can just click on the letter A at the top of the column. You can do this without using the **collapse icon** as long as you can see the letter A.

5) **Repeat the procedure above for the female data:** Position your cursor in the cell where you want to compute the mean for the female word counts, and repeat the procedure above.

6) **Create the Median for Male Word Count:** Move your cursor to the cell for **Male Word Count Median**. In the **Formulas** tab, click on the **Insert Function icon** in the **Function Library group**. In the dialog box that appears, type the word **Median** in the "Search for function" box and click on **GO.** You should see the function highlighted in the "Select a function" box. Click **OK.** Again, in the **Number1** box, select the column, or type in the range of cells where the data for Male Word Counts can be found. Then click on **OK.** If you clicked on the column heading for the data for males, and your data for females is in a contiguous column, you can merely copy the formula from the cell representing Male Word Count Median to the cell representing Female Word Count Median. Notice that your formula will be automatically updated to reflect the fact that the female data is in the next column.

7) **Create the Mode:** Move your cursor to the cell for **Male Word Count Mode**. Follow the same procedures as above, but type **Mode** in your "Search for function" box. Notice that after you have completed entering the cells where the data for Male Word Counts are found, and have clicked on **OK,** the symbol #N/A appears. This indicates that there are no values which repeat in the male (or female) data. **NOTE:** Unfortunately Excel will produce only one mode, even if the data is multi-modal.

8) **Your Table:** Your table should now look something like the one shown. You may have more or less decimal places. To set the number of decimal places shown, select the cells that contain values, then click on the **Home** tab, and click on either the **Increase Decimal icon** or the **Decrease Decimal icon** in the **Number group.**

	Male Word Count	Female Word Count
Mean	15669	16215
Median	14290.14	15916.91
Mode	#N/A	#N/A

9) **Create the Midranges:** You may also be interested in computing the Midrange. Excel does not have a direct way to compute the midrange; however, you can easily create a formula that will return the midrange.

    a. **Determine the Maximum and Minimum values for each column:** Use the **Insert Function icon** in the **Function Library group** of the **Formulas** tab, and first type in **Minimum** in the "Search for function" box, then press **Go.** Choose the option **Min** and as above, choose your column where the data is listed. Then repeat the process but search for **Maximum.** You will choose **Max.** You should see the values listed. Notice that there are decimal answers, even though the data seems to be given in whole numbers. This indicates that the values in the table we used for this manual had been adjusted so that no decimal places were shown.

Minimum	694.7	1673.8
Maximum	47015.6	40054.6

You should select the data in your table, and use the **Decrease decimal icon** to decrease the number of decimal places so that your answers show up as whole numbers.

b. **Create a formula:** Add another row to your table to indicate Midrange. In the slot where the Midrange for Male Word Count would appear, type in a formula similar to: =(F5+F6)/2.

Notice that this formula would indicate that the Minimum Male Word Count is in cell F5, and that the Maximum Male Word Count is in cell F6. If your data appears in different cells, your formula should reflect that.

E	F	G
	Male Word Count	Female Word Count
Mean	15669	16215
Median	14290.1409	15916.9079
Mode	#N/A	#N/A
Minimum	695	1674
Maximum	47016	40055
Midrange	23855	20864

c. **Copy your formula to the other cells:** You can copy the formula in your Male Word Count Midrange to the cell for Female Word Count Midrange. Notice that the relative addresses are updated when you copy your formula. Your table should now look similar to the one shown.

10) **Creating a formula for the Range:** Once you have the Minimum and Maximum values for a data set, you can easily create a formula for the Range. You would set up the formula as follows: = (cell address where the maximum value is listed) – (cell address where the minimum cell address is listed). For the table above, our formula would be: =f6 – f5 for the male range and = g6 – g5 for the female range. When you type the formulas into Excel, you can use either small case or capital letters for the column name.

11) **Finding the Variance and Standard Deviation:** In section 3-3, measures of variation are introduced. Two of the most important measures of variance are the standard deviation and the variance. If you want to include a line for the **sample standard deviation** of your data, you would access the function **STDEV**, and again, select the cells containing the data for which you wanted to compute the sample standard deviation. Likewise, if you wanted the **sample variance**, you would access the function **VAR**, and select the cells containing the data for which you wanted to compute the sample variance. **NOTE:** Excel also allows you to compute the standard deviation and variance of a population, by selecting **STDEVP** and **VARP.**

E	F	G	H
	Male Word Count	Female Word Count	
Mean	15668.5	16215.0	
Median	14290.1	15916.9	
Mode	#N/A	#N/A	
Minimum	694.7	1673.8	
Maximum	47015.6	40054.6	
Midrange	23855.2	20864.2	
Range	46320.9	38380.8	
Std. Dev	8632.5	7301.2	

12) **Creating an Updated Table:** Add the range and standard deviation to our previous table, and then use the **increase or decrease decimal icon** to create values with one decimal place to match the values shown in table 3-3 of your book. Your finished table should appear similar to the one shown.

RANGE RULE OF THUMB

Once we know the mean and the standard deviation of a set of data, we can use the range rule of thumb to find minimum and maximum "usual" values for the data set. Keeping in mind that the majority of values in a data set will lie within two standard deviations of the mean, we can easily set up formulas in Excel to compute these values for us.

1) **Creating the formula for Minimum Usual Value:** Extending the table above, we can set up row titles of "Minimum Usual Value" and "Maximum Usual Value". To compute the Minimum Usual

Value for Male Word Counts, we want to set up a formula that computes the Mean – 2 (Standard Deviation). According to the worksheet where the above table was created, the Mean Male Word Count is in cell F2, and the Standard Deviation for the Male Word Count is in cell F9. Therefore the formula would be: =F2-2*F9. (If the values needed were in different cells in your own worksheet, you would clearly need to create a formula that reflected that.) Since we have used relative addresses, and since the corresponding Mean for Female Word Count and Standard Deviation for Female Word Count are in cells G3 and G9, we can just copy the formula for men, and paste it into the appropriate cell of the table for women. The cell addresses will be automatically updated to reflect the formula: =G2-2*G9

E	F	G
	Male Word Count	**Female Word Count**
Mean	15668.5	16215.0
Median	14290.1	15916.9
Mode	#N/A	#N/A
Minimum	694.7	1673.8
Maximum	47015.6	40054.6
Midrange	23855.2	20864.2
Range	46320.9	38380.8
Std. Dev	8632.5	7301.2
Minimum Usual Value	-1596.55512	1612.58858
Maximum Usual Value	32933.6234	30817.4759

2) **Creating the formula for Maximum Usual Value:**
We will follow a similar process as the previous step, but our formula will contain addition rather than subtraction. Our formula for the Maximum Usual Value for Male Word Counts would be typed in as: =F2 + 2*F9 where again, F2 is the cell containing the mean and cell F9 contains the standard deviation. Again, you can copy this formula over for the Maximum Usual Value for Female Word Counts, and the relative cell addresses will be automatically updated. Your final table should be similar to the one shown.

TO PRACTICE THESE SKILLS

You can apply the skills learned in this section by working on the "Basic Skills and Concepts" exercises found after section 3-2 and 3-3 in your textbook. **Notice that the data for the problems in Section 3-2 mirrors the data that is used again in the problems in Section 3-3.**

> **To practice finding Measures of Center and Measures of Variation from data sets:** Work on the exercises 5 through 24 in both Sections 3-2 and 3-3 Basic Skills and Concepts in your textbook.

- For exercises 5 through 23, you will need to type the data into Excel.

- For exercises 25 and 27, you can load the data from the CD that comes with your book, or from the associated web site. As always, save your work with a file name that is indicative of the problem that you were working on.

SECTION 3-4: MEASURES OF RELATIVE STANDING

DETERMINING HOW MANY UNUSUAL VALUES EXIST IN A DATA SET

When looking at a set of data, it is often useful to know how many values are "unusual". In the last section we learned how to use Excel to find the Mean and Standard Deviation, and then created formulas that allowed us to use these values to create the Minimum and Maximum Usual Values. If we want to figure out how many data values fall outside these boundaries, we can follow the steps below. We will again work with the data from **Data Set 8 in Appendix B** (Word Counts by Males and Females). We will work specifically with the data on Females

1) **Compute the Mean, Standard Deviation and Minimum and Maximum Usual Values for Females:** You can review detailed steps for this in Section 3-2 and 3-3 under the section *Creating Particular Sample Statistics Using the Insert Function Icon* as well as under the section *Range Rule of Thumb.* If you did not save the reorganized data for Word Counts from earlier work, refer back to the instructions under *Opening and Organizing Your Data* at the beginning of section 3-2 & 3-3 to see how to combine the various columns of data in the original data set into just 2 columns.

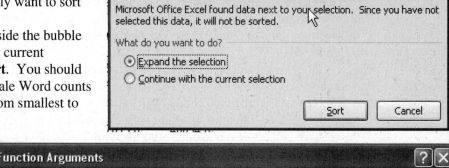

2) **Sort the data in the column for Female Word Counts:** Click on the letter at the top of the column containing Female Word Counts. Then click on the **Sort and Filter icon** in the **Editing group** in the **Home** tab. Choose the option to sort from smallest to largest. A Sort Warning Dialog Box will open. Since you only want to sort the data in the column for Female Word Count, click inside the bubble in front of "Continue with the current selection", and then press **Sort**. You should now see that the data for Female Word counts is listed in numerical order from smallest to largest.

3) **Count the number of Values less than the Minimum Usual Value:** Place your cursor in a blank cell in your worksheet, and then click on the **Insert Function icon** in the **Function Library group** of the **Formulas** tab. In the "Search for function" box, type in **countif** and click on **OK**. This function counts the number of

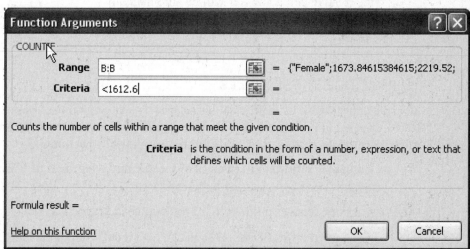

values in a selected range which meet a certain condition. In the Dialog box, select your range by clicking on the letter of the column containing your data for Female Word Counts. In the Criteria box, type in < 1612.6, since this is the minimum usual value we computed using the Range Rule of Thumb. Then click on OK. You should see a value of 0, which makes sense, since in your ordered list, you can see that the minimum value is 1674, and therefore there are no values less than 1612.6.

4) **Count the number of values greater than the Maximum Usual Value:** Use the Countif function again, but this time, in the Criteria box, type in > 30817.5 since this is the Maximum Usual Vlaue we computed using the Range Rule of Thumb. You should see a value of 8, indicating that 8 of your data values are unusual.

COMPUTING STANDARD SCORES OR Z-SCORES

One way to determine if a score is unusual is to compute the z (or standardized) score for a particular value. This tells you how many standard deviations a particular data value is away from the mean. If the z score is less than -2 or greater than 2, we could consider that particular value unusual. We will again work with the data from **Data Set 8 in Appendix B** (Word Counts by Males and Females). We will work specifically with the data on Females

1) **Compute the Mean and Standard Deviation for Female Word Counts:** You can review detailed steps for this in Section 3-2 and 3-3 under the section ***Creating Particular Sample Statistics Using the Insert Function Icon.*** If you did not save the reorganized data for Word Counts from earlier work, refer back to the instructions under ***Opening and Organizing Your Data*** at the beginning of section 3-2 & 3-3 to see how to combine the various columns of data in the original data set into just 2 columns.

2) **Sort the data in the column for Female Word Counts:** See step 2 in the previous section for full instructions on how to sort your data.

3) **Compute the z – score for a set of data values:** From previous work, we already know that the maximum usual value for Female Word Counts is 30817.5. Looking at the sorted list, we can see that there are 8 values larger than this. We will compute the z-scores for these 8 values.

   a. **Set up your initial formula:** Position your cursor in the cell right next to the value of 31,327 in your sorted data. Then click on the **Insert Function icon** in the **Function Library group** of the **Formulas** tab. In the "Search for function" box, type in the word **"standardize"** and then click on **OK**. For each of the categories requested, first click inside the entry box so that you see the blinking line segment in that box.

      i. **For the X value:** Since we want to be able to copy this formula, you should select the cell where your value can be found, or type in the cell address. This will be a relative address that will get updated when we copy the formula.

      ii. **For the mean and standard deviation:** You will need to either type in the **absolute address** for where your values are located, or type in the actual numerical values. This is important in order to have these values remain constant for all computations. Since previous work for this manual had listed the value of the mean in cell G2, and the Standard Deviation in cell G9, we would then want to type in $G$2 and $G$9 so that these cells stayed fixed when we copied the formula. If your values were in different cells, you would use an appropriate address. Click on OK to see the z-score of 2.07 (rounded to 2 decimal places).

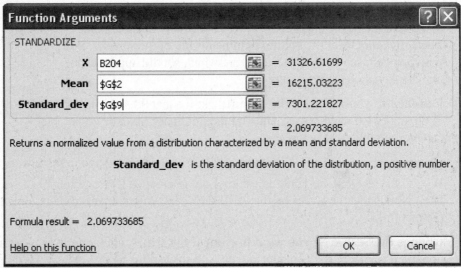

   b. **Format the Standard Score:** Since z-scores are normally reported to only two decimal places, you should format your value so that only 2 decimal places are listed. You can do this by clicking on the cell where your value is listed, and then using the decrease decimals **icon**.

c. **Copy your formula to the remaining cells:** It is easiest to use the fill handle to copy the formula you created into the remaining 7 cells. You should see the values shown. Notice that the largest value in the data set has a z-score of 3.27. This means that the value 40055 is 3.27 standard deviations above the mean. You have already learned that the "usual" values tend to lie within 2 standard deviations of the mean, so therefore this is clearly an unusual value.

31327	2.07
31553	2.10
32291	2.20
33789	2.41
34869	2.55
38154	3.00
39681	3.21
40055	3.27

5- NUMBER SUMMARY AND BOXPLOTS

Excel is not designed to generate boxplots. You can use the **DDXL** Add-In that is supplied with your book to generate this type of graph. Instructions for loading DDXL can be found in section 2-1 of this manual.

1) **Open a worksheet where you have the data on Male and Female Word Counts in two columns:** Refer back to the instructions under *Opening and Organizing Your Data* at the beginning of section 3-2 & 3-3 to see how to combine the various columns of data in the original data set into just 2 columns.

2) **Select the Column of Data you want to work with:** We will create the Boxplot for the data on Female Word Counts. Click on the letter at the top of the column containing this data to select the entire column.

3) **Access DDXL:** Click on the **Add-Ins** tab, and look for DDXL in the **Menu Commands group.** (If you have not previously loaded DDXL, see instructions in section 2-1 of this manual.) Click on the arrow near DDXL and select **Charts and Plots.**

4) **Select Boxplots:** In the Charts and Plots Dialog Box, select Boxplot.

5) **Select the Quantitative Variable:** Under the Names and Columns heading, click on Male. You should notice that the arrow to the right of the Quantitative Variable Box turned blue. Click on this arrow. Since the first row of the column is the variable name, make sure the box is checked next to "First row is variable name". Then click on **OK** at the bottom right hand side of the screen. You will then be taken to the DDXL screen, which should appear as the one displayed on the next page.

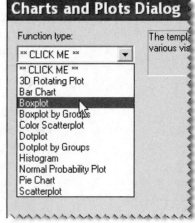

8) **Familiarize yourself with the output:** Notice that the box plot is in the upper left hand corner, and that the Summary Statistics box can be found directly below that.

9) **Create a larger picture:** Click and hold on the diamond shape in the lower right hand corner of the Box plot screen ⬦ and drag this corner out to create a larger graph if you would like. Notice that the scales on the vertical (Y) axis might change when you enlarge the picture.

10) **Change the Scales:** **If you want to control the scales, c**lick on the triangle in the upper left corner of the Box plot screen, and select **Plot Scale.** Make **the changes shown**, and then click on **OK.**

11) **Print or Copy the Boxplot Window:** Make sure you have the Boxplot window selected by clicking anywhere inside the window. You can then use the File, Print command to print the boxplot, or the Edit, Copy Window command to copy the window. Once copied, you can paste the window into another document.

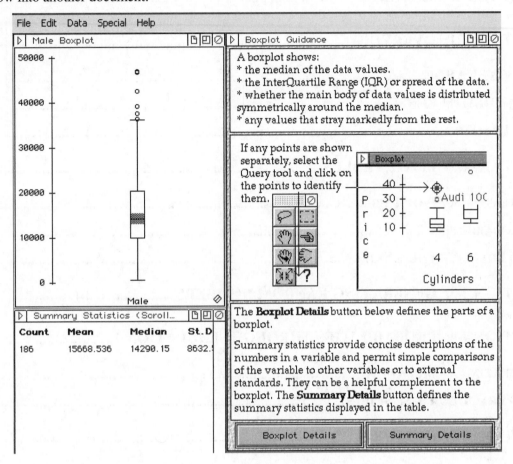

12) **Enlarge the Summary Statistics Window:** Click anywhere in the Summary Statistics window to activate that screen. Again, click and hold on the diamond ⬦ in the lower right hand corner of this box, and drag to the right to expand the amount of the box that can be seen. You want to be able to see all the scores shown in the chart on the next page. Notice that different programs can produce slightly different values for the different percentiles. Though you may see some inconsistency in the exact values returned between programs, any value you get should be in the same general ballpark! Again, once you have selected the window, you can use the Edit, Copy Window to copy the window, and can then paste it into another document.

Count	Mean	Median	St.Dev.	Variance	Range	Min	Max	IQR	25th%	75th%
186	15668.536	14290.15	8632.546	74520858.386	46320.897	694.703	47015.6	10555.3	10009.2	20564.5

TO PRACTICE THESE SKILLS

You can apply the technology skills covered in this section by working through the exercises 31 through 34 from Section 3-4 Basic Skills and Concepts of your textbook. Remember that you may have already loaded some data from the CD into an Excel workbook for work from a previous section. You can open this file, and create a new worksheet within the file for any additional work you do with this particular data set.

CHAPTER 4: PROBABILITY

SECTION 4-1: OVERVIEW

Chapter 4 of your textbook covers the basic definitions and concepts of probability. Many of the concepts presented in this chapter are straightforward and can be done without the use of technology. However there are some features of Excel that you can use when you are working through the material found in this chapter. The following list contains an overview of the topics and functions that will be introduced within this chapter.

PIVOT TABLE
The PivotTable feature is used to help organize, analyze and present summary data. The PivotChart is a helpful tool for visualizing the summary data. It makes it easy to see comparisons, patterns and trends. Both a PivotTable and a PivotChart report allow you to make informed decisions about critical data.

RANDBETWEEN
This function creates columns of random numbers that fall between the numbers you specify. A new random number is returned every time the worksheet is calculated. The function appears in the following format: **RANDBETWEEN (bottom, top)** where bottom is the smallest integer RANDBETWEEN will return and top is the largest integer RANDBETWEEN will return.

FACT
This function returns the factorial of a number. You may recall that in mathematics, the **factorial** of a non-negative number, denoted by $n!$, is the product of all positive integers less than or equal to n. The factorial of a number n is equal to $1 \bullet 2 \bullet 3 \bullet ... \bullet n$. The function appears in the following format: **FACT(number)** where number refers to the nonnegative number for which you want the factorial. If the number is not an integer, it is truncated.

PERMUT
This function returns the number of permutations for a given number of objects that can be selected from a larger group of objects. A permutation is any set or subset of objects or events where internal order is significant. The function appears in the following format: **PERMUT(number, number_chosen)** where number is an integer that describes the number of objects and number_chosen is an integer that describes the number of objects in each permutation.

COMBIN
This function returns the number of combinations for a given number of items. Use COMBIN to determine the total possible number of groups for a given number of items. The function appears in the following format: **COMBIN (number, number_chosen)** where number is the number of items and number_chosen is the number of items in each combination.

SECTION 4-2: PIVOT TABLES

Section 4 – 2 of your textbook introduces Basic Concepts of Probability. You should read through this chapter carefully as it lays the foundation for a number of topics throughout this course. Since this section does not require the use of Excel we will use this opportunity to introduce the concept of a Pivot Table in this section.

In Excel it is possible to generate a table, called a **pivot table,** which can be used to summarize both quantitative and qualitative variables contained within a database. Instead of analyzing rows upon rows of information, a **pivot table** can combine your data and provide you with new ways to look at things. A **pivot table** allows us to create subgroups (or samples), and gives us the ability to find sums, counts, averages, standard deviation and variance of both a sample and population.

CREATING A PIVOT TABLE:

A pivot table can be thought of a summary table of your original data. It can be customized to sort data in a way that meets your individual needs. When you create your pivot table you will begin by defining which fields you wish to view and how the data should be displayed. Based on your specific field selections, Excel combines and organizes the data so you see it in a variety of different views.

We will create a pivot table using the data from Data Set 23: Home Sales in Appendix B in the back of your textbook. This data can also be found on the CD that came with your book. The file name is HOMES.XLS. You can also download this data set at www.aw-bc.com/triola .

We begin by opening Excel and the data file. Each column heading can be thought of as a data field. In this case the data fields are

Selling price	List price	Area
Acres	Age	Taxes
Rooms	Bedrooms	Bathrooms

Using a pivot table allows us to organize and group the data in various ways.

To create a Pivot Table:

There are a couple of things you will want to do before you begin. It is important that each column has a heading, since this will be carried over to the Field List. You will also want to make sure that your cells are properly formatted for their data type.

Excel provides a **PivotTable Wizard** similar to the ChartWizard you have used to generate tables and graphs.

1) Highlight your data

2) Click on the **Insert** tab.

3) Click on the **Pivot Table** icon in the **Tables group** which is located on the left hand side of the **Insert** ribbon.

4) The **Create PivotTable** dialog box will open.

   a. In the **Table/Range** section you will want to check to make sure that you have captured your entire set of data. In this case it should show Sheet1!$A$1:$I$41.

   b. Select the radio button for **New Worksheet** to indicate where you wish your Pivot Table to be placed.

   c. Click **OK**.

5) A new worksheet opens with a blank pivot table. You will notice a **Pivot Table Tools** tab at the top of your Excel worksheet. This ribbon contains a variety of options that you may want to explore once you are comfortable creating a pivot table.

6) You will see a screen similar to the one shown to the right. In addition to the blank pivot table you will see that the fields, which correspond to the column headings, from our original data were carried over to the **PivotTable Field List**. This is located on the right hand side of your worksheet.

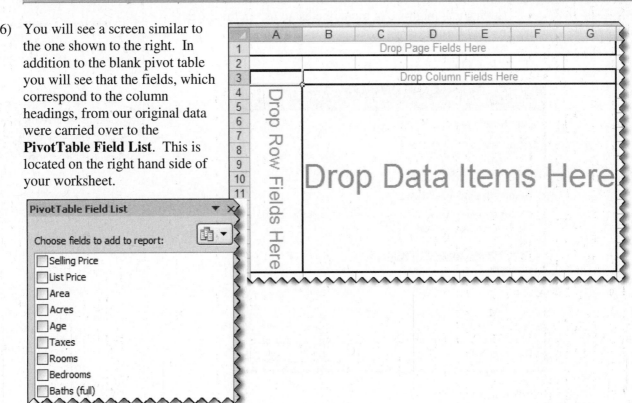

You can now sort your data to create a pivot table that is specific to the information you are interested in gathering.

7) The following steps are designed to help you to create your first pivot table. However you should try different combinations of data in the various fields after you have finished with this first pivot table.

   a. Drag an item such as BEDROOMS from the **PivotTable Field List** to the **Drop Row Fields Here**. You can also drag it down to the **Row Labels** box located on the lower portion of the **PivotTable Field List.** The left side of your Excel spreadsheet should show a row for each total area of the homes for sale. You should also see a checkmark next to SELLING PRICE.

   b. The next step is to ask what you would like to know about each area. Drag ROOMS field from the **PivotTable Field List** to the **Drop Column Fields Here.** You can also drag it down to the **Column Labels** box located on the lower portion of the **PivotTable Field List.** This will provide an additional column for each selling price.

c. To see the count for each selling price, we need to drag the same field to the **Drop Data Fields Here** or to the **Values** box as we have in the columns. We drag ROOMS to this part of the Pivot Table .

d. Notice that Excel determines that we are interested a Sum of Rooms. We can change the **Field Setting** to something else. **Double click** on Sum of Rooms. This will open a menu of choices. Click on **Value Field Settings.**

e. This opens a dialog box which allows us to choose a Field Setting that might be of more interest to us. In this case choose **Count**.

f. Click **OK.**

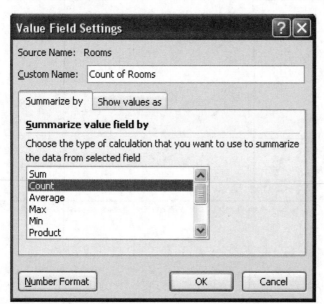

8) Here is our pivot table.

Count of Rooms	Rooms								
Bedrooms	4	5	6	7	8	9	10	11	Grand Total
2	1		3		1				5
3		1	7	6	3	2			19
4					6	4	3	2	15
5							1		1
Grand Total	1	1	10	6	10	6	4	2	40

The table you created shows the number of total rooms and number of bedrooms in each of the forty houses in our data set. For example there was only 1 house in our data set that had 4 rooms and 2 bedrooms. A total of 6 houses in our data had 8 rooms and 4 bedrooms. Only 1 home had 5 bedrooms and that home had 10 rooms total.

As you build your pivot table, you might consider other ways to group the information. For instance you might be interested in knowing how selling price correlates to total are of the house and the number of full bathrooms. You might consider how the price is impacted by the number of bedrooms and full bathrooms. You can play with the various fields by dragging them to the table.

One area that is different is the pivot table has its own options. You can access these options by right-clicking a cell within and selecting **PivotTable Options…** For example, you might only want Grand Totals for columns and not rows.

There are also ways to filter the data using the controls next to Row Labels or Column labels on the pivot table. You may also drag fields to the **Report Filter** quadrant.

Notes:
Once you have created a Pivot table you can refine it. Right click on any cell within the Pivot Table to access the **PivotTable Options** ribbon. As always, we encourage you to explore and experiment with the different options available to you.

Sometimes when you move around your pivot table the **PivotTable Field List** disappears. To get it back just click on any cell inside your pivot table.

PROBABILITIES

It is easier to determine probability information from large sets of data using a pivot table than it would be to sort through the information presented to you in the original Excel worksheet.

TO PRACTICE THESE SKILLS

Use the information presented in the following Excel worksheet to create a Pivot Table that will display the total number of credits broken down by Religion and Major for
   a) male and female students combined
   b) only female students
   c) only male students

	A	B	C	D	E	F
1	SEX	AGE	MAJOR	CREDITS	GPA	RELIGION
2	M	22	Liberal Arts	19	2.5	Jewish
3	M	23	Computer Science	14	3.7	Protestant
4	F	19	Criminal Justice	17	3.8	Protestant
5	M	22	Mathematics	18	2.4	Protestant
6	F	21	English	13	2.5	Jewish
7	F	23	Liberal Arts	18	3	Catholic
8	F	22	Liberal Arts	17	3.2	Catholic
9	M	22	Liberal Arts	18	3.6	Protestant
10	M	22	Education	13	3.5	Catholic
11	F	21	English	17	2.7	Protestant
12	F	22	Criminal Justice	15	2.5	Catholic
13	M	23	Computer Science	15	3.9	Jewish
14	F	22	Computer Science	17	3.3	Jewish
15	M	21	Engineering	12	2.5	Protestant
16	F	22	Engineering	18	4	Protestant
17	F	21	Mathematics	16	3.6	Catholic
18	F	22	Liberal Arts	18	3.4	Jewish
19	M	19	English	17	2.9	Jewish
20	M	21	Criminal Justice	17	3	Catholic
21	F	20	Engineering	16	2.8	Catholic

SECTION 4-3: GENERATING RANDOM NUMBERS

There may be instances where you are interested in setting up an Excel worksheet to do some statistical analysis but you have not collected any real data to test your worksheet on. Generating random numbers in Excel is one way to create a set of data that you can use to test various functions or to create a chart.

There are several functions in Excel that can be used to generate a series of random values. The RAND function always returns a decimal value greater than 0 and less than 1. The RANDBETWEEN function returns a random number between any two values that you specify. We will use the RANDBETWEEN function in this section.

Begin by making cell A1 your active cell.

1) To generate a random set of numbers in Excel, click on the **Formula** tab.

    a. Locate the **Insert Function** icon on the left side of the **Formula** ribbon in the **Function Library** group.

    OR

    b. If you know where your function is located or if you have used it recently you can click on the appropriate icon in the **Function Library** which is also located on the **Formula** ribbon.

2) For this example we will use the first option. Click on the **Insert Function** icon.   The **Insert Function** dialog box will open.

    a. **Click on the down arrow** ▼ by "select a category". Highlight **Math & Trig** as seen on the right.

    b. A new dialog box will open listing a variety of **Math & Trig** functions.

    c. Scroll through the list of function names and highlight **RANDBETWEEN.**

    d. Click **OK**.

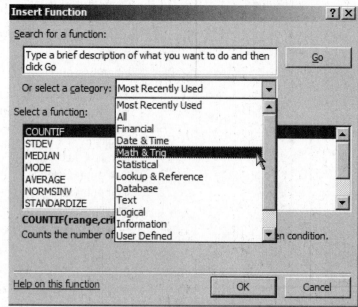

3) A **Functions Argument** dialog box (see below) will open. You are asked to identify the lowest and highest values between which the random number will fall. For example, if you wish to generate a list of possible values that can occur when you roll a single die then the bottom number would be 1 and the top value would be 6. Click **OK.**

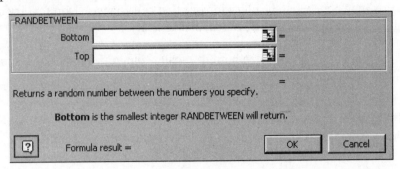

A random number between 1 and 6 should now appear in cell A1. To generate 25 such numbers copy (by dragging) this cell entry to cell A25. When you have completed this task you should have 25 entries in the first column, each between 1 and 6. A new random number is returned every time the worksheet is calculated.

SECTION 4-4: PROBABILITIES THROUGH SIMULATION

The material in this section corresponds to Section 4 – 6 in your textbook.

The goal of every statistical study is to collect data and to use that data to make a decision. Often collecting data or repeating a trial a large number of times can be impractical. With the use of technology we can often simulate an event.

Your textbook defines a **simulation** as follows: *"A simulation of a procedure is a process that behaves the same way as the procedure, so that similar results are produced."*

The random number generator can be used to simulate a variety of statistical problems.

CREATING A SIMULATION

Let's consider the **Gender Selection example** presented in section 4-6 of your textbook. We can simulate the 100 births mentioned in the Example.

1) Begin by opening a new worksheet in Excel. Make cell A1 your active cell.

2) Use the method for generating a random number that was outlined in Section 4 – 3 to simulate 152 births. We choose 0 as the bottom value and 1 as the top value. This will cause the random number generator to display values of 0 or 1. We will let 0 represent a male birth and 1 represent a female birth.

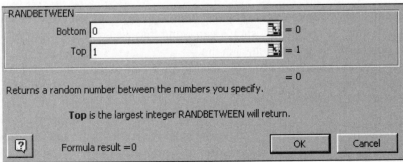

3) To generate 152 random values you can copy the information through to cell A100. You might want to copy the random numbers in such a way so that they are not in one long column. To do this, copy the first random number through to cell A32. You can then copy the information into columns B, C and D. The advantage of this second method is that you can see the 152 entries in your simulation more easily. You may notice that the entry in your first cell change. Don't worry about this. The cells will continue to change until we do our statistical analysis.

4) To determine the probability that the newborn if female, count the number of 1's in your data and divide by the total number of entries.

Simulating Multiple Events

Consider a problem that requires more than one event to occur such as the probability of a specific sum when two dice are tossed.

Suppose that for the purpose of this simulation we wish to find the P (sum of 5) when two dice are tossed 50 times. One possibility would be to toss a pair of dice 50 times and record the sums after each roll. Using technology to simulate this experiment we will produce similar results.

1) Begin a new worksheet with cell A1 as the active cell. Type "First Die" in cell A1, type "Second Die" in cell B1 and type "Sum" in cell C1.

2) **In cell A2** use the method outlined in Section 4 – 3 to generate a *column* of 50 random values between 1 and 6.

3) **In cell B2** generate a *column* of 50 random values between 1 and 6. As before, you may notice that the cell entries in your first column change. This is normal. The cells will continue to change until we do our statistical analysis.

4) **In cell C2** type **=Sum (A2 + B2)** and press **Enter**.

5) **Copy this formula** down through to cell C51. This should give you the sums of the toss of two dice. Once again you will notice that the entries in your first two columns have changed. This is normal. You should be able to see rather easily that the sum of the first two column entries is reflected in the third column.

Now it is possible to determine how many of our simulated tosses of the dice yield a sum of 5, that is, we can find P(sum of 5).

TO PRACTICE THESE SKILLS

You can practice the Excel skills learned in Sections 4-3 and 4-4 of this manual by working through the following problem.

1) Some role-playing games use dice that contain more sides than the traditional six sided dice most of us are familiar with. Assume we are playing such a game and that we are using a pair of ten sided dice which contain the numbers one through ten each die.
   a) Use the Random Number Generator to simulate rolling a pair of ten sided dice fifty times.
   b) Use the results to determine a list of possible sums from the fifty rolls of the dice.
   c) Find the probability of rolling a sum of 20.

2) Develop a simulation using Excel for Exercises 9 and 11 in the Basic Skills and Concepts for Section 4-6.

SECTION 4-5: FACTORIAL

In looking at the solution to the **Chronological Order** example found in section 4-7 of the textbook we see that we are required to multiply $3 \cdot 2 \cdot 1$. This product can be represented by 3! which is read as "three factorial." When reading the solution to the **Routes to National Parks** example we encounter the product $6 \cdot 5 \cdot 4 \cdot 3 \cdot 2 \cdot 1$. This product can be represented by 6! which is read as "six factorial." What appears to be an exclamation point after both the 3 and the 6 is really a **factorial symbol (!).** The factorial symbol indicates that we are to find the product of decreasing positive integers.

In statistics we will encounter a number of formulas that include a factorial symbol somewhere in the formula.

In Excel we can locate the **factorial function** using the same method we used in Section 4 -3 for generating a random number.

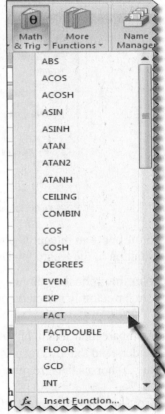

1)   Click on the **Formulas** tab and locate the **Math & Trig** icon in the **Function Library** group.

2) Click on the **Math & Trig** icon

3) Scroll through the function names and **highlight and click on FACT**.

4) When the FACT dialog box opens **enter that number you wish to expand by using factorials**.

    a.   FACT(3) returns a result of 6.

    b.   FACT(6) returns a result of 720

5) Click **OK**

Note: Remember that FACT(6) is equivalent to $6 \cdot 5 \cdot 4 \cdot 3 \cdot 2 \cdot 1$

SECTION 4-6: PERMUTATIONS AND COMBINATIONS:

Problems involving permutations and combinations such as those found in section 4-7 of your textbook can be done fairly quickly and easily using Excel. As noted in your textbook, when we use the term permutation we imply that order is taken into account. This means that different orderings of the same items are counted separately. Permutations are different from combinations, which do not count different arrangements separately.

To access the **PERMUT (permutations)** function and **COMBIN (combinations)** function we click on the **Formulas tab** and make use of the **Function Library** group again.

a.   The use the **PERMUT (permutations)** function we must first access the list of available **Statistical** functions. This is done by clicking on the **More Functions** icon. Scroll down to select **Statistical**. This will display a list of statistics functions. Scroll through the list of function names and **highlight and click on PERMUT.**

b.   The **COMBIN (combinations)** function **can be found in the Math & Trig list of functions.**

Consider the **Exacta Bet** example found in Section 4-7 of your textbook. We will use the **PERMUT function** box to mirror the work done in that example. When you highlight and click on PERMUT as outline above a dialog box will open. You will be asked to supply a number and a number chosen. Each of these is explained below the picture shown here.

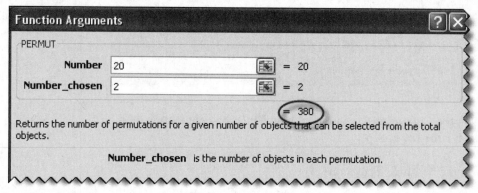

Number – an integer that refers to the total number of objects - in this case there are 20 different horses running in the Kentucky Derby.

Number_chosen - an integer that identifies the number of objects in each permutation. – in this case there are 2 randomly selected horses for the exacta bet which are chosen without replacement.

Compare this result of 380 circled above with the result for this example found in your textbook. We understand that this result tells us that these are 380 different possible arrangements of 2 horses selected from the 20 horses that are in the Kentucky Derby.

Consider the **Phase I of a Clinical Trial example** found in Section 4-7 of your textbook. In this example it is important to notice that we use are using permutations for part (a) of this example because order does count. We are using combinations for part (b) of this problem because order does not count. Use the process outlined using the PERMUT function to complete part (a) of this example.

We will focus on part (b) and use the **COMBIN function** to mirror the work done in the second part of this example. When you highlight and click on COMBIN as outline above a dialog box will open. You will be asked to supply a number and a number chosen as before. Each of these is explained below the picture of the COMBIN function dialog box.

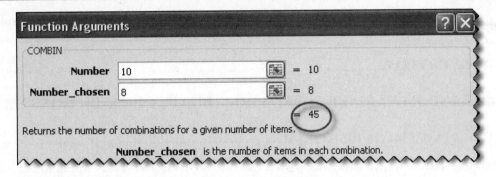

Number - the number of items – in this case there are 10 available people.

Number_chosen - the number of items in each combination – in this case there are 8 people selected.

Compare this result of 45circled above with the result for this example found in your textbook. We understand that this result tells us that these are 45 different possible treatment groups.

If you completed part (a) you would have found that when all of the different possible arrangements are taken into account there are 1,814,400 permutations. When order is not taken into account there are only 45 combinations.

TO PRACTICE THESE SKILLS

You can practice the Excel skills learned in this section of this manual by working through the problems 5 – 12 to become familiar with each of these functions. Then try problems 17, 19, 21, and 27 from the Basic Skills and Concepts for Section 4-7 to extend your knowledge.

CHAPTER 5: DISCRETE PROBABILITY DISTRIBUTIONS

SECTION 5-1: OVERVIEW

Excel has functions built into it that can be used to calculate the probabilities associated with several different probability distributions. Computing these probabilities by hand can be very time consuming. Although tables are available for some distributions, these are also limited in scope. Excel provides you with a tremendous amount of flexibility in creating these distributions quickly and efficiently. The following list contains an overview of the new functions that will be introduced within this chapter.

FILL SERIES
This feature enables us to quickly and efficiently enter a string of consecutive numbers in a column or row.

BINOMDIST
This function returns the individual term binomial distribution probability.

POISSON
This function returns the Poisson probability that a particular number of occurrences of an event will occur over some interval.

SECTION 5-2: RANDOM VARIABLES AND PROBABILITY DISTRIBUTIONS

Let's consider the Example 1 from section 5-2 in your text book. In this case, x is the random variable, and it represents the number of peas with green pods among 5 offspring peas. We will work with the values given, and create a probability histogram, as well as consider how we can use Excel to compute the mean and standard deviation of this probability distribution. In the next section, we will actually learn how to generate the probabilities in this distribution using Excel.

CREATING A PROBABILITY HISTOGRAM

1) **Open a new worksheet.** Again, it is a good idea to get in the habit of entering the data in a worksheet entitled "Original Data." You can then copy the data to another worksheet, and work with the data there, ensuring that you always have ready access to the original data if needed.

2) **Enter the distribution:** Enter the column headings and values given in the table shown.

3) **Copy your data and rename your worksheet:** Select this data, and copy it to another worksheet. Give your new worksheet an appropriate name that will remind you what is there. You might want to call the worksheet something like "Prob Dist." to indicate that you have the probability distribution in this sheet.

	A	B
	x Number of Peas with	
1	Green Pods	P(x)
2	0	0.001
3	1	0.015
4	2	0.088
5	3	0.264
6	4	0.396
7	5	0.237
8		

4) **Access the Column Chart Icon:** To access the **Column Chart icon**, you can click on the **Insert** tab, and then click on **Column** in the **Charts group**. Select the first type of Column Chart in the menu. A blank Chart Area will appear in your worksheet.

5) **Select the Data you want to graph:** Right click somewhere inside the Chart Area, and choose **Select Data**. The **Select Data Source** box will open.

a. **Select the values that represent the vertical axis:** Under the **Legend Entries (Series)** box, click **Add.** The **Edit Series** dialog box will appear. Use the **collapse and expand icons** to help in selecting the appropriate cells for this dialog box. Select the cell where you have the word Probability for the Series Name, and select the cells where the probabilities are for the Series Values. Then click on **OK.**

b. **Select the values that represent the horizontal axis:** Under the **Horizontal (Category) Axis Labels** box, click **Edit.** Using the **collapse icon**, select the cells where your random variables are listed, and after clicking on the **expand icon**, click on **OK.** This takes you back to the **Select Data Source** box, where you should again click on **OK.** You should see the chart below.

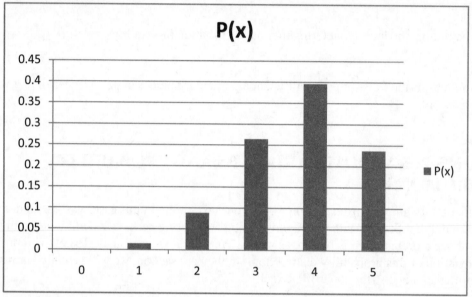

6) **Modify your Chart:** Refer back to **Modifying the Initial Histogram** directions in sections 2-2 & 2-3 of this manual to recall how to:

a. Chose a Design that eliminates the Gap between the bars

b. Create clearer borders for your bars.

c. Change the Graph Title and Horizontal Labels

d. Resize a Region

e. Make other changes

Your final relative frequency histogram should look similar to the one shown on the next page.

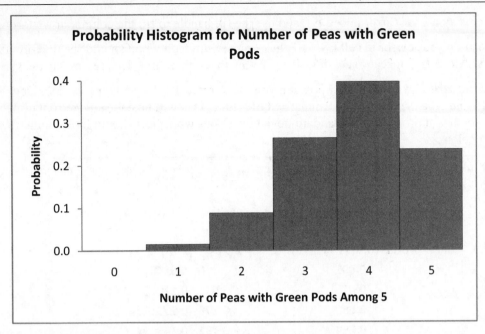

COMPUTING THE MEAN, VARIANCE AND STANDARD DEVIATION

We can compute the mean, variance and standard deviation of a probability distribution according to formulas 5 – 1, 5 – 3 and 5 – 4 in your book. To use Excel to work with these formulas, follow the steps below. The work below assumes that your x values are in column A, starting in cell A2, and that your probabilities are in column B, starting in cell B2. (See the print out on the next page.) If your data is in a different location, you should make adjustments as necessary to the specific cell information shown below.

1) **Create the products of random variables and corresponding probabilities:** In cell D1, type in: x * P(x) as the title for your column. Then in cell D2, type in the following formula: = A2 * B2. This tells Excel to multiply the random variable in cell A2 by its associated probability in cell B2. After pressing **Enter**, you should see a 0 in cell D2, since the x value is 0, and anything multiplied by 0 is 0.

2) **Copy your formula down:** Position your cursor back in cell D2. Then use the fill handle to copy the formula down through cell D7.

3) **Find the mean:** Formula 5-1 tells us that we need to add these products up. Position your cursor in cell C9, and type "Mean". Then position your cursor in cell D9, and click on the **Sum icon** in the **Editing group** of the **Home** tab. To find the sum of the appropriate values, select the cells D2 to D7. Notice that the formula: =SUM(D2:D7) appears in cell D9, and the column of values above this cell are selected. Press **Enter.** You should now see the value which represents the mean of the probability distribution.

4) **Find the Variance:** To find the variance, we will use Formula 5-2. We need to create another column which allows us to compute $(x - \mu)^2 * P(x)$

    a. Position your cursor in cell E1, and type in the formula: (x-m)^2*P(x), where m represents the mean.

    b. Move your cursor to cell E2, and type in the formula: =(A2-$D$9)^2*B2. Notice that the x value and the probability value are relative cell addresses which will be updated when we copy the formula, but that the mean is an absolute address. Press **Enter**. This formula takes the value in cell A2, subtracts the mean from it, squares that result and then multiplies it by the value in cell B2.

a. Reposition your cursor in cell E2, and use the fill handle to fill the column down through E7.

b. Position your cursor in cell E8, and click on the **Sum icon**. You should see the formula: =SUM(E2:E7) appear in cell E8. Press **Enter** to see this sum. This represents the variance.

5) **Find the Standard Deviation:** Now position your cursor in cell C10, and type in "Std. Dev." Move to cell D 10 and type the following formula: =SQRT(E8). This will take the square root of the variance, which will produce your standard deviation. Your worksheet should look similar to the one shown below:

	A	B	C ↓	D	E	F
1	x Number of Peas with Green Pods	P(x)		x*P(x)	(x-m)^2*P(x)	
2	0	0.001		0	0.014078	
3	1	0.015		0.015	0.113603	
4	2	0.088		0.176	0.270116	
5	3	0.264		0.792	0.149293	
6	4	0.396		1.584	0.024356	
7	5	0.237		1.185	0.369128	
8					0.940574	
9			Mean	3.752		
10			Std. Dev	0.969832		
11						

IDENTIFYING UNUSUAL RESULTS WITH THE RANGE RULE OF THUMB

The range rule of thumb tells us that most values should lie within 2 standard deviations of the mean. If we have computed the mean and standard deviation in Excel, we can easily create formulas which allow us to determine the minimum and maximum usual values.

1) **Creating the minimum usual value:** In the worksheet from above, type the word Minimum Usual Value in cell C12. (You can use the **Wrap Text icon** to create cells like the ones shown below.) Move to cell D12 and type in the following formula: =D9-2*D10. This formula subtracts 2 times the standard deviation from the mean. Press **Enter** to see the result.

2) **Creating the maximum usual value:** Type the word Maximum Usual Value in cell C13. Move to cell D13 and type in the following formula: =D9+2*D10. This formula adds 2 times the standard deviation to the mean. Press **Enter** to see the result. Because of the context of this problem, we would interpret this to mean that the number of green pods among 5 offspring is between 2 and 5. Notice that we use whole numbers, since you can't have 1.8 pods. Therefore, it would be unusual to get just 1 green pod, or no green pods from 5 offspring.

11				
12			Minimum Usual Value	1.812337
13			Maximum Usual Value	5.691663

EXPECTED VALUE

Notice that the value that you computed in your worksheet for the summation of the products of your random variables and their corresponding probabilities can also be called the expected value of a discrete random variable.

TO PRACTICE THESE SKILLS

You can apply these technology skills by working on exercises 7 through 12 from section 5–2 Basic Skills and Concepts in your textbook.

SECTION 5-3: BINOMIAL PROBABILITY DISTRIBUTIONS

Binomial variables take on only two values. One of these values is generally designated as a "success" and the other a "failure." We typically see the probability of a "success" denoted by the letter p and the probability of failure denoted by the letter q. The sum of p and q must equal one, since success or failure are the only possible outcomes.

Suppose we consider a binomial distribution where p = .65 and there are 15 trials. Since there are 15 trials, we know that the random variable can take on the values between 0 and 15 inclusive.

CREATING A COMPLETE DISTRIBUTION

Begin your work in a new worksheet. You may want to name this worksheet something like "Bin. Dist" to indicate that you are creating a binomial distribution.

1) **Create the column of random variables:** First enter a title for the column representing the random variable in cell A1. (If we knew what the random variable represented, we should use a suitable name, otherwise, we will use "x".) Then in cell A2, type in the value 0, since this is the first value of the random variable. Press **Enter**.

2) **Use the Fill Series feature:** Reposition your cursor in cell A2. (Note: In order to activate the fill feature, you **must** move out of the cell where your first value is typed, and then move back to it.) In the **Home** tab, click on the **Fill icon** in the **Editing Group.** Then click on **Series.** You will see the **Series** dialog box open.

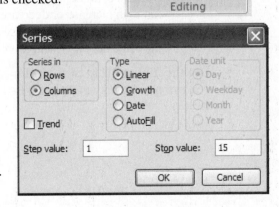

   a. Make sure that the bubble in front of **Columns** is checked.

   b. Make sure that **Linear** is selected under **Type.**

   c. Make sure that the **Step value** is set at 1.

   d. Type in 15 for the **Stop value**, since there are 15 trials in the experiment.

   e. Click on **OK.** You should now see the whole numbers from 0 to 15 in column A.

3) **Create the column for probabilities:** Move to column B and type "P(x)" in cell B1. Press **Enter**. Your cursor should now be in cell B2.

4) **Access the Binomial Distribution Function:** With cell B2 selected, click on the **Insert Formula icon** in the **Function Library group** in the **Formulas** tab. In the "Search for function:" box, type in "binomial". Click on the function **BINOMDIST** in the "Select a function:" box. Click on **OK.**

5) **Fill in the BINOMDIST Dialog Box:** You need to complete the dialog box as follows:

   a. **Number_s** refers to the number of successes. You want to enter the cell address where this information is stored. Since we began our values for x in cell A2, we type in A2.

   b. **Trials** refers to the total number of trials. Type in 15.

   c. **Probability_s** refers to the probability of a success. For this experiment, type in the value .65.

   d. **Cumulative** will list the cumulative probabilities. Since we do not want these at this time, type in "False".

   Click on **OK.** You will now see the probability of getting exactly 0 successes in 15 trials if the probability of success is .65. **Note:** This value may be written in exponential notation. You can reformat this value by clicking on the down arrow in the **Number group** in the **Home** tab, and changing from Scientific to Number. You can then increase or decrease the number of decimal places by using the **increase/decrease decimals icons** to show the numbers with more or less decimal places.

6) **Use the fill handle to copy the formula:** With cell B2 activated, you can use the fill handle to fill in the remaining values in the table.

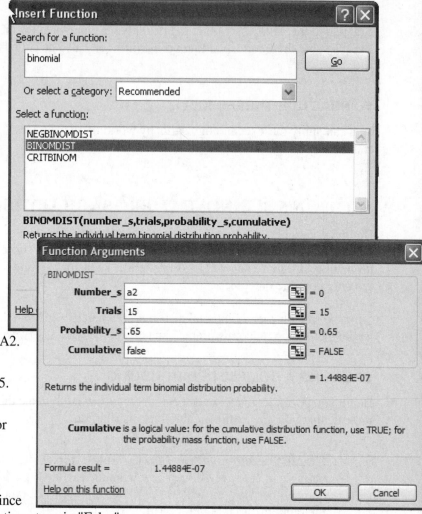

	A	B	C
1	x	P(x)	
2	0	0.0000001449	
3	1	0.0000040361	
4	2	0.0000524687	
5	3	0.0004222484	
6	4	0.0023525267	
7	5	0.0096117520	
8	6	0.0297506611	
9	7	0.0710372928	
10	8	0.1319264010	
11	9	0.1905603570	
12	10	0.2123386835	
13	11	0.1792469406	
14	12	0.1109623918	
15	13	0.0475553108	
16	14	0.0126167151	
17	15	0.0015620695	
18			

CREATING A WORKSHEET THAT WILL COMPUTE A SPECIFIC PROBABILITY

You may not need to find the entire probability distribution, but may just want to create a worksheet which would make it easy to plug in appropriate values and find a particular probability.   Follow the steps below to create such a worksheet.

1) **Give your worksheet a title:** Create a clear title for your worksheet. Notice that we called our worksheet "Worksheet to Compute Specific Binomial Probabilities". Since this name took up more than one cell, it's a good idea to select the cells that the title spans and then click on the **Merge icon** in the **Alignment group** in the **Home** tab.

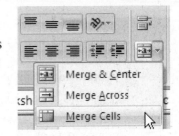

2) **Set up the categories you need to input:** Type the specifics of the information that you would need to enter into Excel. You need to enter the values for n, p and x.

3) **Format the background for the cells where you will be typing in values:** To make it clear where you will be typing in your values, you may find it helpful to format those cells with some background color. Position your cursor in a cell where you will need to enter a value. Then click on the **Fill color icon** in the **Font group** of the **Home** tab. You can either choose one of the colors there, or you can click on the "More Colors" option and select a color from that dialog box. You can then copy this formatting to the other cells by pressing and holding the **Ctrl** key followed by the **C** key on your keyboard. Move your cursor to the next cell you want to have this background color and press and hold the **Ctrl** key followed by the **V** key on your keyboard. To turn off the copy command (notice that you see a blinking frame around the cell you copied), press **Esc.**

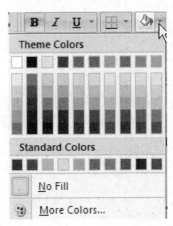

4) **Input specific values in the cells:** We will let n = 6, p = .75 and x = 4.

5) **Apply the BINOMDIST function to these values:** With your cursor in cell E9 for the worksheet shown, access the Binomial Distribution Function by clicking on the **Insert Formula icon** in the **Function Library group** in the **Formulas** tab. In the "Search for function:" box, type in "binomial". Click on the function **BINOMDIST** in the "Select a function:" box. Click on **OK.**

	A	B	C	D	E	
1	Worksheet to Compute Specific binomial probabilities					
2						
3	Enter the number of successes (x)				4	
4						
5	Enter the sample size (n)				6	
6						
7	Enter the probability of success (p)				0.75	
8						
9	The value of P(x) is:				0.296631	
10						

6) **Fill in the BINOMDIST Dialog Box:** You need to complete the dialog box by typing in the cell address where each value is located. Then click on **OK**. Your final worksheet should look like the one shown.

TO PRACTICE THESE SKILLS

You can apply these technology skills by working on the following exercises. Make sure you save your work using a file name that is indicative of the material contained in your worksheets.

1) Read exercise 31 from Section 5-3 Basic Skills and Concepts in your textbook. Use Excel to set up the **probability distribution** for this exercise. Before beginning your work in Excel, make sure you clearly identify what the values for your random are, and what value you should use for the probability of "success." You can use Excel to compute the sums of appropriate cells from your probability distribution. For example, to find the probability of **at least 12** of the people selected have brown eyes, you want to include the probabilities for x = 12, 13 and 14. You can use the **Sum icon** to add up these probabilities.

2) Read exercises 34 and 35 from Section 5-3 Basic Skills and Concepts in your textbook. Use Excel to set up the **probability distribution** for this situation. Use your **Sum icon** to help add up the appropriate values from the table to answer the questions asked.

3) You can use the worksheet you created for computing specific binomial probabilities for exercises 15 through 24 from section 5-3 Basic Skills and Concepts in your textbook.

SECTION 5-4: MEAN, VARIANCE, AND STANDARD DEVIATION FOR BINOMIAL DISTRIBUTION

Although you can use the formulas presented in section 5-2 to compute the mean, variance and standard deviation for the binomial distribution, there are easier formulas to work with for this particular distribution.

- You can find the mean by multiplying the sample size (n) by the probability of success (p).

- You can find the standard deviation by taking the square root of the product formed by multiplying the sample size (n) by the probability of success (p) and the probability of failure (q).

- You should use formulas preceded by the "=" sign when you enter your information into Excel.

Suppose we used the values: n = 14, p = 0.5 and q = 0.5. You can follow the steps below to create a table for this binomial experiment.

1) **Give your worksheet a title:** Create a clear title for your worksheet. Notice that we called our worksheet "Mean and Standard Deviation for a Binomial Probability". Since this name took up more than one cell, it's a good idea to merge the cells that the title spans. Use the **Merge icon** in the **Alignment group** in the **Home** tab.

2) **Set up the categories you need to input:** In our worksheet, we typed the information "Sample Size", "Probability of Success", "Probability of Failure", "Mean", and "Std. Dev" in cells A2 through A6. You can expand the size of column A by positioning your cursor on the right hand side of the box containing the letter A until you see a double headed arrow appear. Then double click to have Excel automatically adjust the width of the column to fit the contents of the cells.

3) **Format the background for the cells where you will be typing in values:** To make it clear where you will be typing in your values, you may find it helpful to format those cells with some background color. Position your cursor in a cell where you will need to enter a value. Then click on the **Fill color icon** in the **Font group** of the **Home** tab. Once you have selected your fill color, you can then copy this formatting to the other cells by pressing and holding the **Ctrl** key followed by the

C key on your keyboard. Move your cursor to the next cell you want to have this background color and press and hold the **Ctrl** key followed by the **V** key on your keyboard. To turn off the copy command (notice that you see a blinking frame around the cell you copied), press **Esc.**

4) **Type in your specific values:** We will use an example where n is 14, and p is .5. In cell B2 type in 14. In cells B3 type in 0.5.

5) **Set up a formula to compute q:** Since q can be computed by subtracting p from 1, in cell B4, type in =1-B3. When you press **Enter**, you should see the value shown in the table below.

6) **Set up the formula to compute the mean:** Since the mean of a binomial probability distribution is n * p, in cell B5, type in the formula: =b2*b3. Then press **Enter.**

7) **Set up the formula to compute the standard deviation:** Since the standard deviation of a binomial probability distribution is $\sqrt{n \cdot p \cdot q}$, in cell B6 type in the formula: =sqrt(b2*b3*b4). Then press **Enter.**

8) **Your final table:** Your table should look similar to the one shown below:

	A	B	C
1	**Mean and Standard Deviation for a Binomial Probability**		
2	Sample Size (n)	14	
3	Probability of Success (p)	0.5	
4	Probability of Failure (q)	0.5	
5	Mean	7	
6	Std. Dev	1.870828693	
7			

TO PRACTICE THESE SKILLS

Once you have the table from above set up in a worksheet, you can change the values for sample size, probability of success and probability of failure. Your values for the mean and standard deviation should be automatically updated. You can copy this table, paste it in other cells of your worksheet, and then change the numbers in the copy of the table for a different problem.

1) Create a table similar to the one above, but using the data from exercise 9 from Section 5-4 Basic Skills and Concepts in your textbook. To complete part b, you want to compute the values that are 2 standard deviations above and below the mean. Add lines to your table which will use the numbers generated to compute the minimum usual value ($\mu - 2\sigma$) and the maximum usual value ($\mu + 2\sigma$). You may find it helpful to refer back to section 5-2 of this manual under Identifying Unusual Results with the Range Rule of Thumb.

2) You can now make additional copies of your table, and paste them to other parts of your worksheet if you want to keep a separate record for each problem, or you can merely update the information in the one table, and just keep track of your answers on separate paper. You can work on exercises 10 through 20 from Section 5-4 Basic Skills and Concepts in your textbook.

SECTION 5-5: CREATING A POISSON DISTRIBUTION

In a Poisson distribution, the random variable x is the number of occurrences of the event in an interval. The interval can be time, distance, area, volume, or some similar unit.

Using the example on **Earthquakes** from section 5-5 in your text, we can generate a table to find the probabilities asked for in the example.

1) **Create a column for the random variable:** In cell A1, type in "x". Then type the values 0, 1, 2, 3, 4, 5, 6 and 7 in that column. (You could use the **Series** option available in the **Fill icon** in the **Editing group** of the **Home** tab to accomplish this.)

2) **Create a column for the probabilities:** In cell B1, type in "P(x)".

3) **Access the POISSON Function:** Position your cursor in cell B2. Then click on the **Insert Function icon** in the **Formulas** tab. In the "Search for function:" box, type in "Poisson". You will see the function **POISSON** highlighted in the "Select a function:" box. Click on **OK.**

4) **Fill in the POISSON Dialog Box:** You should fill in the dialog box as follows:

   a. For the **X** input, type in cell A2, since this is where your first random variable is located.

   b. In the input box following **Mean**, type in .93 since this is the computed mean for this example.

   c. In the **Cumulative** box, type in "**False**".

5) Click on **OK**. You will now see the probability that there were no earthquakes in a randomly selected year. You can change the number of decimal places for your answer by using the **increase/decrease decimals icons**.

6) **Use the fill handle to fill in the remaining cells:** With your cursor on cell B2, move your mouse until you see the black plus sign in the lower right hand corner. Holding your left mouse button down, drag this down to fill the rest of the column.

7) **Set up a column for the Expected Number of Earthquakes:** Suppose we want to use this probability to compute the Expected Number of Earthquakes. In cell C1, type in "Expected Number of Earthquakes". To have the text "wrap" to fit in to the formatted cell width, click on the **Wrap Text icon**.

8) **Compute the Expected Number of Regions:** Move to cell C2. We want to multiply each probability by the total number of years (100). Enter the formula: =100*b2 into cell C2. Then use the fill handle to copy this formula down into the remaining cells. Again, to match the style of the table in the book, you can format your column to show 1 decimal place. You will notice that some of the numbers generated are slightly different from those in the table in the book. This is because even when Excel is displaying only 3 decimal digits based on our cell formatting, it is using the longer string in computations.

9) **Finishing touches:** To make your table easier to read, you can select the entire table, and then press the **centering icon** in your tool bar. ![centering icon]. Your table should now look like the one shown.

	A	B↓	C
1	x	P(x)	Expected number of Earthquakes
2	0	0.3945537	39.45537
3	1	0.3669350	36.69350
4	2	0.1706248	17.06248
5	3	0.0528937	5.28937
6	4	0.0122978	1.22978
7	5	0.0022874	0.22874
8	6	0.0003545	0.03545
9	7	0.0000471	0.00471
10			

TO PRACTICE THESE SKILLS

You can apply these technology skills by working on the following exercises. Make sure you save your work using a file name that is indicative of the material contained in your worksheets.

1) Create a table showing "x" and "P(x)" for exercise 9 from Section 5-5 Basic Skills and Concepts in your textbook.

2) Create a table showing "x", "P(x)" and the "Expected Number" columns for exercise 10 from Section 5-5 Basic Skills and Concepts in your textbook.

CHAPTER 6: NORMAL PROBABILITY DISTRIBUTIONS

SECTION 6-1: OVERVIEW

In this chapter we will explore how to compute probabilities for a normal distribution, as well as find specific values if we are given information about the probability for a particular normal distribution. There are essentially five different functions that we can use when exploring normal distributions.

NORMSDIST: This function returns the **standard** normal cumulative distribution for a specified z value.

NORMSINV: This function returns the inverse of the standard normal cumulative distribution for a specified z value.

STANDARDIZE: This function returns a standardized score value for specified values of the random variable, mean and standard deviation.

NORMDIST: This function returns the cumulative normal distribution for specified values of the random variable, mean and standard deviation.

NORMINV: This function returns the inverse of the normal cumulative distribution for specified values for the probability, mean and standard deviation.

SECTION 6-2: WORKING WITH THE STANDARD NORMAL DISTRIBUTION

The first normal distribution presented in your text is the standard normal distribution. This distribution has a mean of 0 and a standard deviation of 1.

In this section of the manual, we will set up worksheets which will allow us to enter information for the different situations that might arise for computing probabilities and finding values. We can then simply enter values into appropriate cells of the worksheets to find the answers requested.

FINDING PROBABILITIES GIVEN VALUES

In order to use Excel to find probabilities, we must recognize that the program computes probabilities by determining the total area under the normal distribution **from the left up to a vertical line at a specific value**. This is vital for us to remember as we work to set up appropriate formulas.

1) **Give your worksheet a title:** Create a clear title for your worksheet. Notice that we called our worksheet "Worksheet to Compute Probabilities for the Standard Normal Distribution". Since this name took up more than one cell, it's a good idea to select the cells that the title spans and then merge them using the **Merge & Center icon** found in the **Alignment group** in the **Home** tab.

2) **Set up the categories you need to input:** We need to create a worksheet which would allow us to enter in one boundary value, and then compute either P (z < a) or P (z > a). We also need to set up a situation where we could enter in a lower and upper value and compute P (a < z < b). In order to find this probability, we need to generate the values P (z < b) and P (z < a) in Excel. Use the worksheet shown below to set up appropriate categories.

3) **Format the background for important cells:** To make it clear what the various parts of your worksheet represent, and where you will be typing in your values, you may find it helpful to format those cells with some background color. Position your cursor in a cell that you want to stand out.

Then click on the down arrow near the **Fill Color icon** in the **Font group** of the **Home** tab. You can either choose one of the colors there, or you can click on **More Colors** and select a color from that dialog box You can then copy this formatting to the other cells by pressing and holding the **Ctrl** key followed by the **C** key on your keyboard. Move your cursor to the next cell you want to have this background color and press and hold the **Ctrl** key followed by the **V** key on your keyboard. To turn off the copy command (notice that you see a blinking frame around the cell you copied), press **Esc.** You can use a variety of colors to help clarify your worksheet.

	A	B	C	D	E	F
1			Worksheet to Compute Probabilities for a Standard Normal Distribution			
2						
3	Given one value					
4	Enter the Value	P(z < value)	P(z > value)			
5						
6						
7		Between Two Values				
8	Enter Lower Value	Enter Upper Value	P(z < Lower Value)	P(z < Upper Value)	P(Lower < z < Upper)	
9						

4) **Input specific values in the cells:** Let's use a value of -1.23 for the one value situation, and let's use values of -2 and 1.5 for the two value situation. If you set your worksheet up like the one shown above, you should enter -1.23 in cell A5, -2 in cell A9, and 1.5 in cell B9.

5) **Set up the formula for P (z < value) by accessing the NORMSDIST function:** Remember, Excel will always return a value which represents the area to the left of a particular value. Position your cursor in cell B5 and click on **Insert Function icon** in the **Functions Library group** of the **Formulas** tab. (Notice the symbol for the **Insert Function icon** can also be found at the beginning of the Formula bar. You may activate this dialog box by clicking on this icon in the formula bar.) In the "Search for function:" box, type in "standard normal distribution". You will see the function

NORMSDIST highlighted in the "Select a function:" box. (The **S** indicates that this is the Standard Normal Distribution.) **Click** on **OK.** Fill in the dialog box using the appropriate cell reference for where your value is located. For the worksheet we set up, we would fill our dialog box out as the one shown. Then click on **OK**.

Function Arguments

NORMSDIST

z A5   = -1.23

= 0.109348552

Returns the standard normal cumulative distribution (has a mean of zero and a standard deviation of one).

Z is the value for which you want the distribution.

Formula result =   0.109348552

Help on this function   OK   Cancel

6) **Copy and update this formula to the other cells:** Rather than accessing the dialog box twice more for the remaining cells where you want P (z < value), we can copy the formula we set up in cell B5 to the other cells. **Always make sure that you have appropriate cell references after you have copied the formula to a new location!**

a. **Position your cursor in cell B5:** Press and hold **Ctrl** and then press **C** on your key board. You should now see a blinking border around B5.

b. **Move your cursor to cell C9:** Press and hold **Ctrl** and then press **V** on your key board. Your formula should have been copied to this new cell.

c. **Check the formula line at the top of the page:** Always make sure you check that when you copied the formula, you are referring to the appropriate cells. In this case, we want P (z < Lower). Since the lower value is in cell A9, we need to adjust the formula. Position your cursor up in the Formula bar right in front of the letter B. Delete B and type in A, since your lower value is in cell A9. (Remember, it doesn't matter whether you use lower or upper case letters in Excel when referencing cell addresses.)

d. **Move your cursor to cell D9:** Press and hold down **Ctrl** and then press **V** on your keyboard. This will copy the formula again.

e. **Check the formula line at the top of the page**. Notice that you want P (z < Upper), and since the upper value is in cell B9, everything is ok. **The main part of your sheet should look like the one shown:**

	A	B	C	D	E
4	Enter the Value	P(z < value)	P(z > value)		
5	-1.23	0.109348552			
6					
7	Between Two Values				
8	Enter Lower Value	Enter Upper Value	P(z < Lower Value)	P(z < Upper Value)	P(Lower < z < Upper)
9	-2	1.5	0.022750132	0.933192799	
10					

7) **Create formulas for the remaining cells:** Since Excel's NORMSDIST function will only return the area under the normal curve to the left of the value we use, we need to create formulas that use this information to help determine P (z > value) and P (Lower < z < Upper).

a. **P (z > value):** Since the area under the entire normal curve is 1, if we want P (z > value), we can use the formula 1 – P (z < value). Position your cursor in cell C5, and type in the formula =1-b5. Then press **Enter**. You should now see 0.890651448 in cell C5.

b. **P (Lower < z < Upper):** Since the area we want is between the upper and lower value, we can create the formula which computes P (z < Upper) – P (z < Lower). Position your cursor in cell E9 and type in the formula: =d9-c9. Then press **Enter.** You should now see 0.910442667 in cell E9.

8) **Try some other values:** You can now just type in other values in the shaded cells, and your spreadsheet will automatically update the values for the probabilities. Check that your worksheet is set up correctly by typing the values shown in the yellow cells in the worksheet below. Your values should be updated to show the values for the probabilities below.

	A	B	C	D	E
1	Worksheet to Compute Probabilities for a Standard Normal Distribution				
2					
3	**Given one value**				
4	Enter the Value	P(z < value)	P(z > value)		
5	1.25	0.894350226	0.105649774		
6					
7	Between Two Values				
8	Enter Lower Value	Enter Upper Value	P(z < Lower Value)	P(z < Upper Value)	P(Lower < z < Upper)
9	1	2	0.841344746	0.977249868	0.135905122
10					

FINDING A SCORE WHEN GIVEN THE PROBABILITY

Let's assume that we are working with thermometers that are normally distributed with a mean of 0 degrees Celsius and a standard deviation of 1 degree Celsius. Suppose we are given a probability and we want to find the value that corresponds to that probability. A classic example of this would be if we wanted to find the value for a particular percentile. Remember, in an earlier chapter you learned that P_{80} represents the value such that at least 80% of the values would be less than or equal to that value and at least 20% of the values were greater than or equal to that value.

Again, let's work to set up a worksheet which will allow us to enter appropriate values into certain cells, and which then would automatically compute the values we wanted to know. We want to create a worksheet that looks like the one below.

	A	B	C	D	E
1	**Finding Values from Known Areas**				
2	If you have the area to the LEFT of the value you want:	Enter the area as a decimal	Enter the mean	Enter the Std. Dev	Value to 2 decimal places
3		0.8	0	1	0.84
4					
5	If you have the area to the RIGHT of the value you want:	Enter the area as a decimal	Enter the mean	Enter the Std. Dev	Value to 2 decimal places
6		0.2	0	1	0.84

1) **Set up your column headings:** Set up a chart like the one shown on the previous page, making use of the **Wrap Text icon** to have the titles fit as shown.

2) **Format the background color for the cells where we will enter information:** Again, there are certain cells where we will enter information for the situation. While we know the mean and the standard deviation of a standard normal distribution remain at 0 and 1 respectively, we are creating a sheet that can also be used for other normal distributions. Notice that two situations are set up: if you're given the area to the **Left** of the value you want, and if you're given the area to the **Right** of the value you want.

3) **Access the NORMINV Function to find the desired value where area is to the LEFT of the value:**

   a. **Position your cursor in cell E3**, and click on the **Function icon** on the Formula bar (or click on the **Insert Function icon** in the **Formulas** tab.). In the "Search for function:" box, type in "inverse of normal distribution". You will see the function **NORMINV** highlighted in the "Select a function:" box. Click on **OK.**

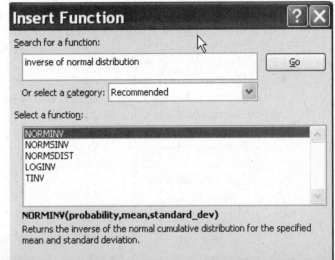

   b. **Fill in the dialog box**: You will use the cell reference where the area is located for **Probability.** You should also use the appropriate cell references for where the **Mean** and **Standard Deviation** are found. If you set your worksheet up like the one shown, your dialog box should look like the one shown. **Keep in mind that Excel will always interpret the probability you put in as being to the LEFT of the value you want.**

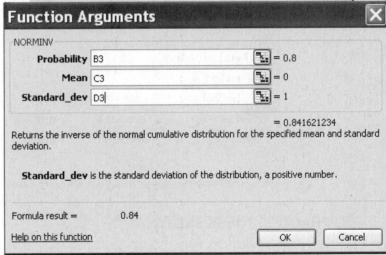

4) **Find the value where area is to the RIGHT of the value:** We can again copy the formula we have entered above, and paste it into the cell which represents the value for area to the **RIGHT**. However, we will have to modify the formula.

   a. **Position your cursor in cell E3:** Press and hold **Ctrl** and then press **C** on your key board. You should now see a blinking border around E3.

   b. **Move your cursor to cell E6:** Press and hold **Ctrl** and then press **V** on your key board. Your formula should have been copied to this new cell.

   c. **Check the formula line at the top of the page:** Always make sure you check that when you copied the formula, you are using appropriate values. Keeping in mind that Excel always understands that Probability is the area to the LEFT of the value you want, we need to make a minor change to our copied formula. Your copied formula should look like:

=NORMINV (B6,C6,D6). We need to realize that since cell B6 contains the area (or probability) to the RIGHT of the value we want, we need to adjust the formula so that we see =NORMINV(1-B6,C6,D6). Position your cursor in the formula line at the top of the worksheet right before the B and type in 1 - . Then press **Enter.** You should now see .84 in cell E6. This should make sense. If the value .84 separates the bottom 80% of values, it should be the same score which separates the top 20% of values.

5) **Try some other values:** To make sure that your worksheet is set up correctly, enter an area to the **LEFT** of .025, and an area to the **RIGHT** of .025. Keep your mean and standard deviation at 0 and 1 respectively. The values shown in the table should make sense. If you have an area of .025 to the LEFT of the value you want, the value will have to be to the left of the mean, and since this is the standard normal distribution, it must therefore be negative. If you have an area of .025 to the RIGHT of the value you want, the value will have to be to the right of the mean. Since this is the standard normal distribution, the value must therefore be positive. Since you have equal areas in the "tails", it makes sense that the values are opposites of each other for the standard normal distribution.

	A	B	C	D	E	F
2	If you have the area to the LEFT of the value you want:	Enter the area as a decimal	Enter the mean	Enter the Std. Dev	Value to 2 decimal places	
3		0.25	0	1	-0.67	
4						
5	If you have the area to the RIGHT of the value you want:	Enter the area as a decimal	Enter the mean	Enter the Std. Dev	Value to 2 decimal places	
6		0.25	0	1	0.67	
7						

TO PRACTICE THESE SKILLS

1) You can apply the skills you learned in this section to find probabilities by using Excel to complete exercises 17 through 36 from Section 6-2 Basic Skills and Concepts in your textbook.

2) You can apply the skills you learned in this section to find scores associated with particular percentiles or probabilities by using Excel to complete exercises 41 through 52 from Section 6-2 Basic Skills and Concepts in your textbook.

SECTION 6-3: APPLICATIONS OF NORMAL DISTRIBUTIONS

Your textbook presents the idea that if we are working with a normal distribution which is not a standard normal, we could opt to "standardize" the scores and then use the techniques for finding probabilities and values as presented in the Standard Normal section.

Since Excel allows us to work directly with non-standard normal distributions, we will modify our previous spreadsheet entitled "Worksheet to Compute Probabilities for the Standard Normal Distribution". (See section 6-2.)

FINDING PROBABILITIES USING THE NORMDIST FUNCTION

If you completed and saved your work from section 6-2, you can modify that worksheet rather than creating an entirely new one. You would need to change the title, and insert a couple of more columns for the additional information we need for a non-standard normal distribution. You will also have to change from the NORMSDIST function to the NORMDIST function. If you feel comfortable with this, simply copy and paste your standard normal distribution worksheet to a new worksheet, and begin making appropriate modifications. You will want to make sure you review the step where we introduce the NORMDIST function. Also, notice that we no longer use "z" to indicate a value, we use "x". The letter "z" is specifically reserved for the standard normal distribution.

You may want to just begin again, so that you can practice setting up worksheets. The directions below take you through each step.

1) **Give your worksheet a title:** Create a clear title for your worksheet. Notice that we called our worksheet "Worksheet to Compute Probabilities for a Normal Distribution". Since this name took up more than one cell, it's a good idea to select the cells that the title spans and then click on the **Merge & Center icon** in the **Home** tab.

2) **Set up the categories you need to input:** We need to create a worksheet which would allow us to enter in one boundary value, and then compute either $P(x < a)$ or $P(x > a)$. We also need to set up a situation where we could enter in a lower and upper value and compute $P(a < x < b)$. In order to find this probability, we need to generate the values $P(x < b)$ and $P(x < a)$ in Excel. Use the worksheet shown below to set up appropriate categories.

3) **Format the background for the cells where you will be typing in values:** To make it clear what you are given, and what the areas of the worksheet represent, you may find it helpful to format those cells with some background color. Position your cursor in a cell where you will need to enter a value. Then click on the drop down arrow by the **Fill Color icon** in the **Home** tab. Choose the color you want for the background fill in the cell. You can then copy this formatting to the other cells by pressing and holding the **Ctrl** key followed by the **C** key on your keyboard. Move your cursor to the next cell you want to have this background color and press and hold the **Ctrl** key followed by the **V** key on your keyboard. To turn off the copy command (notice that you see a blinking frame around the cell you copied), press **Esc.**

	A	B	C	D	E	F
1	Worksheet to Compute Probabilities for a Normal Distribution					
2						
3	Given One Value				P(x < Value)	P(x > Value)
4	Enter the Value	Enter the Mean	Enter the Std. Dev.			
5						
6						
7	Between Two Values					
8	Enter the Lower Value	Enter the Upper Value	Enter the Mean	Enter the Std. Dev.	P(x < Value)	
9					P(x > Value)	
10					P(Lower < x < Upper)	

4) **Input specific values in the cells:** Let's use data which assumes that weights of men are normally distributed with a mean of 172 pounds and a standard deviation of 29 pounds. Let's consider the probability that a randomly selected man will weigh less than 174 pounds. Let's also consider the probability that a randomly selected man will weigh between 150 and 170 pounds.

a. **Enter the One Value:** We want to find the probability that a randomly selected man will weigh less than 174 pounds, so we will enter 174 in cell A5, the mean (172) in cell B5 and the Standard Deviation (29) in cell C5.

b. **Enter the Two Values:** We want to find the probability that a randomly selected man will weigh between 150 and 170 pounds. We will enter 150 in cell A9, 170 in B9, the mean (172) in C9 and the Standard Deviation (29) in D9.

5) **Set up the formula for P (x < value) by accessing the NORMDIST function:** Remember, Excel will always return a value which represents the area to the **left** of a particular value. Position your cursor in cell E5 and click on the **Insert Function icon** in the Formula bar. In the "Search for function:" box, type in "normal distribution". You will see two normal distribution functions listed. **Make sure you highlight the function NORMDIST in the "Select a function:" box.** Click on **OK.** Fill in the dialog box using the

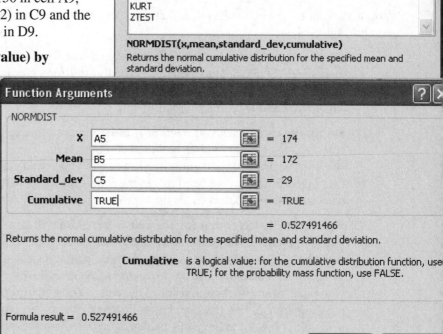

appropriate cell references for where your values are located. For the worksheet we set up, we would fill our dialog box out as the one shown. Then click on **OK**.

6) **Copy and update this formula to the other cells:** Rather than accessing the dialog box again for the remaining cells where you want P (z < value), we can copy the formula we set up in cell E5 to the other cells. **Always make sure that you have appropriate cell references after you have copied the formula to a new location!**

a. **Position your cursor in cell E5:** Press and hold **Ctrl** and then press **C** on your key board. You should now see a blinking border around E5.

b. **Move your cursor to cell F8:** Press and hold **Ctrl** and then press **V** on your key board. Your formula should have been copied to this new cell, however you see a screen that looks like:

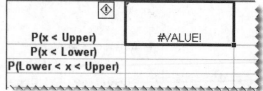

c. **Check the formula line at the top of the page:** Always make sure you check that when you copied the formula, you are referring to the appropriate cells. In this case, we want P (x < Upper). Notice that the formula at the top of the page shows: =NORMDIST(B8,C8,D8,TRUE). Notice that because of the way we set up our worksheet, we actually want B9, C9 and D9. Simply position your cursor in the formula line and make appropriate changes so that you see: =NORMDIST(B9,C9,D9,TRUE). Now press **ENTER.**

d. **Move your cursor to cell F9:** Press and hold down **Ctrl** and then press **V** on your keyboard. This will copy the formula again.

e. **Check the formula line at the top of the page.** Notice that you want P (x < Lower), but the formula bar shows cell B9, whereas you want the first value to be A9. Position your cursor up in the formula line right in front of the letter B. Delete B and type in A, since your lower value is in cell A9. Your formula should now look like: =NORMDIST(A9,C9,D9,TRUE). Press **Enter.**

7) **Creating the remaining formulas:** We still need values for P (x > value) and P (Lower < x < Upper).

a. **P (x > value):** Position your cursor in cell F5. Since the total area under the curve is 1, to find the P (x > value) we just need to create the formula: 1 – P (x < value). In cell F5, type in the formula: =1-E5 and then press **Enter.**

b. **P(Lower < x < Upper):** Position your cursor in cell F10. Since we can find the value of the area bounded by the lower and upper boundaries by subtracting P (x < Upper) – P (x < Lower), in cell F10 type the formula: =f8-f9 and then press **Enter.**

c. **Your sheet should look like the one shown:**

	A	B	C	D	E	F
1	Worksheet to Compute Probabilities for a Normal Distribution					
2						
3	Given One Value				P(x < Value)	P(x > Value)
4	Enter the Value	Enter the Mean	Enter the Std. Dev.			
5	174	172	29		0.527491466	0.472508534
6						
7	Between Two Values					
8	Enter the Lower Value	Enter the Upper Value	Enter the Mean	Enter the Std. Dev.	P(x < Upper)	0.472508534
9	150	170	172	29	P(x < Lower)	0.224039746
10					P(Lower < x < Upper)	0.248468787

FINDING VALUES FROM KNOWN AREAS

Refer back to the worksheet that we created in section 6-2 for "Finding the Score When Given a Probability". We set this worksheet up so that it could be used with any normal distribution. Simply enter appropriate information for the exercise involved into this worksheet. For a non-standard normal, our mean and standard deviation do not have to be 0 and 1 respectively.

TO PRACTICE THESE SKILLS

Once you have the tables above set up, you can quickly change the values that you input. Excel will automatically update the values in the columns containing the formulas. To ensure that you can always retrieve the original table that you set up by following the instructions above, it is always best to copy the table to another worksheet before you begin modifying it.

After you complete an exercise, you may want to copy and paste your completed table to another location in your worksheet. After you have copied the table, you should activate the cell where you want your table to begin. Then click on **Edit, Paste Special**, and click in the bubble by the word **Values**. Then click on **OK**. If you activate one of the cells where you had entered a formula originally, you will notice that now only the value shows up. The original formula is no longer active in the copied table. You can go back to the original table to compute the value for the next problem.

1) Try using the table you set up to find Probabilities using the NORMDIST function for exercises 5 through 8 and 13 through 16 from section 6-3 Basic Skills and Concepts in your textbook.

2) Try using the table you set up to find Values from Known Areas, and modify the numbers as appropriate to address exercises 9 through 12 and exercises 18 through 20 from Section 6-3 Basic Skills and Concepts in your textbook.

3) Work with the appropriate table, and modify the numbers as appropriate to address exercises 21 through 34 from Section 6-3 Basic Skills and Concepts in your textbook.

SECTION 6-4: SAMPLING DISTRIBUTIONS AND ESTIMATORS

This manual does not contain any new material specifically associated with section 6-4. You may find it helpful to review the material on finding the mean from a probability distribution presented back in section 5-2 of this manual.

SECTION 6-5: THE CENTRAL LIMIT THEOREM

In this section, we will use Excel to help us visualize the Central Limit Theorem. We can also modify the worksheet we created earlier to compute probabilities for a normal distribution so that we can compute probabilities for situations where the Central Limit Theorem needs to be used.

VISUALIZING THE CENTRAL LIMIT THEOREM

To help us see some of the major concepts behind the Central Limit Theorem, we will generate a table containing 1500 randomly generated digits 0, 1, 2, 3, ..., 9. We will set the worksheet up in 30 columns and 50 rows.

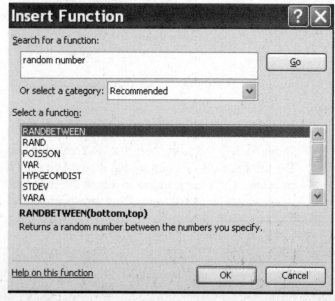

1) **Name your worksheet:** In your worksheet, type "Randomly Generated Digits" in cell A1.

2) **Use the Random Number Generator:** Position your cursor in cell A2, and click on the

Function icon on the Formula bar (or click on the **Insert Function icon** in the **Formulas** tab.) type Random Number in the "Search for Function" box, and press **Enter**. You should see RANDBETWEEN listed in the "Select a Function" box. Make sure this is highlighted, and press **ENTER**. In the dialog box, enter 0 for the Bottom value and enter 9 for the Top value. Then click on **OK**.

3) **Use the fill handle to create our table of values:** We want to create 30 columns of randomly selected values, where each column has 50 rows. Make sure your cursor is positioned in cell A2, and using the fill handle, pull your mouse horizontally until you are out in column AD. Then release your mouse. Access the fill handle again, and pull down vertically until you are in row 51. You should now have a table of 1500 randomly generated digits. Your table should look different from others, in that the values in each cell are being randomly generated. Just be aware of this if you are comparing your table to another classmate's table.

4) **Find the mean of each row:** Position your cursor in cell AF1 and type in "ROW MEAN".

    a. **Access the AVERAGE function:** Position your cursor in cell AF2, and click on the **Function icon** on the Formula bar. In the "Search for a function" box, type in Mean, and press **Enter**. You should see **AVERAGE** highlighted in the "Select a function" box. Click on **OK**. In the input box by Number 1, type in the range "A2:AD2", and click on **OK**. In cell AF2, you should now see the average of the thirty numbers in row 2.

    b. **Use the fill handle to copy the formula:** Position your cursor in cell AF2. Access the fill handle, and pull the mouse down to cell AF51. You should now have the average for each row of 30 values.

5) **Stabilize your table:** You may notice that as you work with your columns and cells, the table of numbers you generated changes. To stabilize the set of data you are working with, highlight the columns containing your values and your means and click on the **Copy icon** from the **Clipboard group** in the **Home** tab. Move to a new worksheet and position your cursor in cell A1. Click on the down arrow under the **Paste icon** in the **Clipboard group** of the **Home** tab, and then click on **Paste Values**. You will now have a stable table, since the cells are representing numbers now rather than formulas. Your pasted table may very well be quite different from the table that you chose to copy! From here out, the copied table will remain stable, as it is no longer associated with the random number generator.

6) **Create a histogram for the 1500 digits found in your table:** Use the Upper Class Limits of 0, 1, 2, 3, ..., 9 so that we get a distribution that shows how many of each digit occurred in the data. Refer back to Chapter 3 as necessary to review how to create a frequency distribution and histogram similar to the one shown below. Remember, since your values were randomly generated, your information does not have to be exactly the same as that shown below. You should see clearly that your histogram does not appear bell shaped.

Digits	Frequency
0	159
1	158
2	146
3	146
4	167
5	147
6	154
7	147
8	146
9	130

Distribution of 1500 Values

7) **Create a histogram for the 50 Sample Means:** To make it easy to compare the pictures, use the

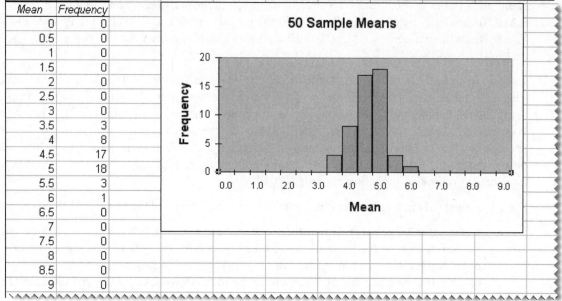

Mean	Frequency
0	0
0.5	0
1	0
1.5	0
2	0
2.5	0
3	0
3.5	3
4	8
4.5	17
5	18
5.5	3
6	1
6.5	0
7	0
7.5	0
8	0
8.5	0
9	0

"bins" or upper class limits from 0 to 9 in intervals of 0.5. Although the histogram of sample means may not look entirely bell shaped, it is definitely moving in that direction.

8) **Computing the Means and Standard Deviations:** In order to see more of the results of the Central Limit Theorem, we want to compute the mean of both the original 1500 values as well as the mean of the 50 sample means. We will also want to compute the Standard Deviations for each set of numbers.

   a. **Use the AVERAGE Function:** Set up an area like the one shown below, and use the AVERAGE function for the entire 1500 values (A2:AD51), as well as for the 50 sample means (AF2:AF51).

   b. **Use the STDEVP Function:** Since we are considering the 1500 values a population, we should use the STDEVP function to find the standard deviation of this population. Apply the function to the values in cells (A2:AD51).

   c. **Use the STDEV Function:** Since we are considering a samples when we compute the sample means, we would use the STDEV function to find the standard deviation of these sample means. (AF2:AF51)

   d. **Computing the Theoretical Standard Deviation:** The Central Limit Theorem says that the standard deviation of the sample means should be the standard deviation of the population divided by the square root of the sample size. Set up a formula that shows the appropriate cell address where your population standard deviation is located, and which looks like the following: = (cell address of population SD)/sqrt(30).

9) **Notice the relationships:** Notice that the mean of your population and the mean of your sample means comes out to be the same (or very, very close), while the standard deviation of your sample means is fairly close to your computed $\sigma / \sqrt{n}$.

Mean of 1500 Values	4.388		Mean of 50 Sample Means	4.388
SD of 1500 Values	2.845134		SD of 50 Sample Means	0.506142
			Theoretical SD	0.519448

COMPUTING PROBABILITIES THAT INVOLVE THE CENTRAL LIMIT THEOREM

We have previously constructed worksheets that allow us to enter particular values and then use appropriate functions in Excel to compute the probabilities associated with these values. We can modify the worksheet we created for computing probabilities for normal distributions (See section 6-3) so that we can use it for situations involving sample means.

1) **Copy and paste your previous worksheet:** Assuming you have previously created and saved the worksheet called: "Worksheet to Compute Probabilities for a Normal Distribution" (see section 6-3), copy and paste that worksheet to another worksheet.

2) **Insert a column for the sample size:** In order to use the Central Limit Theorem, we need to be able to call on the sample size. If your worksheet is set up as the one from section 6-3, click at the top of column E. Then click on the down arrow by the **Insert** command in the **Cell group** in the **Home** tab. From the menu, click on **Insert Sheet Columns.** You should see a column inserted in the worksheet to the left of your probability headings.

3) **Change your column headings:** You will need to make a number of changes to your headings. Change your headings to match the ones shown below. Notice that x changes to x bar to indicate you are talking about the probability of a sample mean. Also, since we don't have any particular values entered yet, the cells that are defined by a formula show: #DIV/0! Don't worry; this will change once we enter values in our shaded cells.

	A	B	C	D	E	F	G
1	Worksheet to Compute Probabilities for the CLT						
2							
3	Given One Value					P(x bar< Value)	P(x bar > Value)
4	Enter sample Value	Enter Population Mean	Enter Population SD	Enter Sample Size			
5						#DIV/0!	#DIV/0!
6							
7	Between Two Values						
8	Enter the Lower Value	Enter the Upper Value	EnterPopulation Mean	Enter Population SD	Enter Sample Size	P(x bar< Upper)	#DIV/0!
9						P(x bar< Lower)	#DIV/0!
10						P(Lower < x bar < Upp	#DIV/0!
11							

4) **Change your formulas:** There are 3 places where we need to adjust our previously entered formulas, cell F5, G8 and G9.

   a. **Cell F5:** We need to change the standard deviation that was originally used when we set the worksheet up for the normal probabilities. We want to use a standard deviation which would be the population standard deviation (cell C5) divided by the square root of the sample size (cell D5). We can put our cursor in cell F5 and modify the formula in the formula bar so that we see: =NORMDIST(A5,B5,C5/SQRT(D5),TRUE). Then press **Enter.**

   b. **Cell G8:** Again, we need to change the standard deviation. Your formula should look like: =NORMDIST(B9,C9,D9/sqrt(E9),TRUE)

    c. **Cell G9:** This needs a similar change to the standard deviation, and should look like:
=NORMDIST(A9,C9,D9/SQRT(E9),TRUE)

5) **Input some specific values:** Let's use the information from your textbook Example 2 on Water Taxi Safety, part b. You are given a sample size of 20. You were told that the Mean of the population was 172 and the standard deviation of the population was 29.

    a. **Suppose we wanted P(xbar > 174):** Enter 174 for your sample value, 172 for your population mean, 29 for your population SD, and 20 as your sample size. You should see that the probability that you would get a sample mean greater than 174 is about .37888. This means that there is approximately a 37.88% chance that the mean weight of 20 randomly selected men would be greater than 174.

    b. **Suppose we wanted P (172 < xbar < 175):** In the second part of the worksheet, enter 172 as your lower value, 175 as your upper value, 172 as your population mean, 29 as your population SD, and 20 as your sample size. You should see that the probability that you would get a sample mean greater than 172 but less than 175 is about .1782. This means that there is approximately a 17.82% chance that the mean weight of 20 randomly selected men would be between 172 and 175.

	A	B	C	D	E	F	G	H
1	Worksheet to Compute Probabilities for the CLT							
2								
3	Given One Value					P(x bar< Value)	P(x bar > Value)	
4	Enter sample Value	Enter Population Mean	Enter Population SD	Enter Sample Size				
5	174	172	29	20		0.621119823	0.378880177	
6								
7	Between Two Values							
8	Enter the Lower Value	Enter the Upper Value	EnterPopulation Mean	Enter Population SD	Enter Sample Size	P(x bar< Upper)	0.678186903	
9	172	175	172	29	20	P(x bar< Lower)	0.5	
10						P(Lower < x bar < Upp	0.178186903	
11								

TO PRACTICE THESE SKILLS

You can apply the skills you learned in this section by working on the following exercises.

1) Repeat the exercise presented in this section, but using a table with 100 rows of 30 digits. Compare the histograms you create from your new data set to those presented in this section.

2) Many of the exercises in Section 6-5 of your textbook deal with finding probabilities. Use the worksheets: "Worksheet to Compute Probabilities for a Normal Distribution" and "Worksheet to Compute Probabilities for the CLT" as appropriate to work on the exercises beginning with # 5 from section 6-5 Basic Skills and Concepts in your textbook

SECTION 6-6: NORMAL DISTRIBUTION AS APPROXIMATION TO BINOMIAL

If we have a binomial distribution where $np \geq 5$ and $nq \geq 5$, we can approximate binomial probability problems by using a normal distribution. You can convince yourself that as the number of trials increases for a binomial distribution, the graphs of the distributions look more and more like a normal distribution.

1) **Create binomial probability distributions for the following scenarios.** Look back to section 5-3 for instructions on how to create the distributions.
- Let the first distribution have n = 10 and p = 0.5
- Let the second distribution have n = 25 and p = 0.5
- Let the third distribution have n = 50 and p = 0.5

2) **Create the probability histograms for your 3 distributions:** Refer back to section 5-2 and sections 2-2 and 2-3 of this manual for help in accomplishing this.

3) **Notice what happens to the graphs:** What you should notice about each of these successive pictures is that the areas in the bars, which represent the probabilities for each random variable, more closely approximates the area contained under the curve. If you think of the curve as representing a normal curve, than you can clearly see that the binomial probability distribution is more closely approaching a normal distribution.

4) **Compute the probabilities:** We can see that the probabilities we find using the normal distribution will closely match those that we can generate from the binomial distribution. To see this clearly, we will use the case of n = 50 and p = 0.5.

    a. **Add the binomial probabilities:** Using the values in our binomial distribution table, we can find $P(x \geq 30)$ by adding up the probabilities from x = 30 to x = 50. This produces a value of .101319.

    b. **Finding cumulative binomial probabilities without the entire table:** Most often, unless we were graphing the entire binomial probability distribution, we would not typically generate the entire list of values. Suppose we just wanted to quickly compute the probability above. In Excel, access the **BINOMDIST**

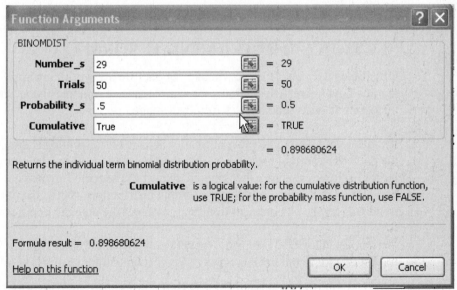

function, and fill out the dialog box as shown. Notice that Cumulative is marked as True. This tells Excel to add up the probabilities for x = 0 to x = 29. Since we want the probability that x is greater than or equal to 30, subtract this value from 1 to find the probability of .101319.

c. **Find the probability using the normal distribution:** If we want to use a normal distribution, we will need to compute the mean and standard deviation using the formulas for a binomial distribution. You should find the value for the mean is 25 and the standard deviation is 3.535534. As outlined in section 6-6 of your textbook, we need to use a continuity correction when using the normal distribution to approximate the binomial. Since we want to find the probability of getting a value greater than or equal to 30, we should find $P(x \geq 29.5)$. Use the **NORMDIST** function on Excel, or use the worksheet you created previously. Remember that Excel returns a value that represents the area under the normal distribution curve to the left of the value. In order to find the probability we really want, we will need to subtract this value from 1. The **NORMDIST** function returns a value of 0.898454. Subtracting this value from 1 produces a probability of .101546. Recall that the probability using our binomial information is .101319. If we were to increase our sample size, we would find that we get even closer approximations when using the normal distribution to approximate the binomial.

TO PRACTICE THESE SKILLS

You can apply the skills you learned in this section by working on the following exercises.

1) Create the binomial probability distribution where n = 100 and p = 0.5, and create the probability histogram. Compare your picture to those for n = 10, 25 and 50 found in this section.

2) Use the **NORMDIST** function on Excel, and the continuity correction to find the probabilities for exercises 13 and 15 in Section 6-6 Basic Skills and Concepts in your textbook. To recall how to work with the NORMDIST function, you may want to refer back to instructions found in Section 6-3 of this manual.

3) Use the **BINOMDIST** function on Excel, and, when appropriate, the **NORMDIST** function with the continuity correction to find the probabilities asked for in the odd numbered exercises 17 - 31 from Section 6-6 Basic Skills and Concepts in your textbook.

SECTION 6-7: DETERMINING NORMALITY

Oftentimes we want to know whether the data we are working with is normally distributed. We have already learned how to create histograms for sample data. From our histogram, we can reject normality if the histogram departs dramatically from bell shape.

An alternate way to determine normality is to construct a normal quantile plot (or normal probability plot) for the data. In a normal probability plot, the observations in the data set need to be ordered from smallest to largest. These values are then plotted against the expected z scores of the observations calculated under the assumption that the data are from a normal distribution. When the data are normally distributed, a linear trend will result. A nonlinear trend suggests that the data are non-normal.

We can use the **DDXL** Add-In to generate a normal probability plot. We will create a model using the data for **Diet Coke** found in Data Set 17: Weights and Volumes of Cola.

1) **Load the data from the CD that comes with your book (COLA.XLS), or from the website:** Copy the column showing the 36 weights of Diet Coke (CKREGWT) into column A of a new Excel worksheet.

2) **Select the column you want to consider:** It is essential that before you access DDXL, you select the column(s) that you want to work with. Since we want to look at the weights of Diet Coke, click on the A at the top of the column.

3) **Access DDXL:** Click on the **Add-Ins** tab, and then click on **DDXL** in the **Menu Commands group.**

4) **Click on Charts and Plots**. Click on the down arrow on the Function type box, and click on **Normal Probability Plot.** In the **Names and Columns** box, click on the name CKDIETWT, and then click on the blue arrow by the Quantitative Variable box to move the data name to that box. Make sure you have the box checked in front of "First Row is Variable Name" if this is and case. Click on **OK.** You should see the following information on your screen.

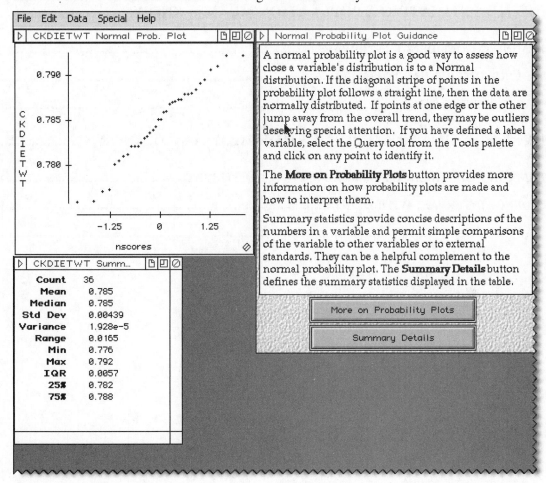

TO PRACTICE THESE SKILLS

You can apply the skills from this section by working with exercises 13 - 16 from section 6-7 Basic Skills and Concepts in your textbook. For each exercise, you should open the file in an Excel workbook from the CD that comes with your book or from the Addison Wesley website (www.aw.com/triola).

CHAPTER 7: ESTIMATE AND SAMPLE SIZES

SECTION 7-1: OVERVIEW

It is not unusual for us to be in a situation where we do not know, or have access to, the statistical parameters that are used to describe a population. When this occurs we need to use information contained in a given sample representative of our population. If we can identify a numerical value that describes a given sample, then this value can also be used to estimate the corresponding statistical descriptor for the population this sample represents. How well the sample statistic estimates our population value is always an issue. A **confidence interval** addresses this issue because it provides a range of values which is likely to contain the population parameter. **Confidence intervals** are important in statistics because they allow you to gauge how accurately a sample parameter approximates the same parameter with respect to the population. The confidence interval consists of a range (or interval) of values instead of just a single value. The confidence interval also contains a probability. This probability value tells you the likelihood that you have an interval that actually contains the value of the unknown population parameter. Components of a confidence interval include an upper and lower limit for the parameter under consideration, as well as a probability value.

Excel does not have a built in function that automatically calculates the confidence interval. Instead, we will need to rely on some of the functions we have already used in Excel to help us build a confidence interval. In addition to creating a confidence interval with Excel we will make use of the add-in DDXL. This add-in was supplied on the CD that came with your textbook. If you do not have the original CD you can find and download DDXL at the website www.aw-bc.com/triola . If you did not load the DDXL add-in or cannot find it on the computer you are working with please go back to Chapter 2, Section 1 and follow the instructions for loading DDXL.

The following list contains an overview of the functions we will be utilizing in this chapter.
CONFIDENCE
Returns the confidence interval for a population mean.

CONFIDENCE (alpha, standard_dev, size)
Alpha refers to the significance level used to compute the confidence interval, standard_dev is the population standard deviation for the data range and size is the sample size.

TINV:
Returns the inverse of the Student's t-distribution for the specified degrees of freedom.

TINV (probability, deg_freedom)
Probability is the probability associated with a two tailed Student's t distribution and deg_freedom is a positive integer indicating the number of degrees of freedom needed to characterize the distribution.

SECTION 7-2: ESTIMATING A POPULATION PROPORTION

As stated in the Overview, Excel does not produce confidence interval estimates for proportions. It will be necessary for us to use the **DDXL** add-in that came as a supplement to your textbook to do this. **If you have not loaded DDXL yet please do so before going any further.** The instructions for adding this feature to Excel can be found in Chapter 2, Section 1 of this manual. If you are unsure as to whether or not you have already added DDXL click on the Add – in tab. If you have already loaded DDXL you will see it in the **Menus Command group** which is located on the left hand side of the **Add – in** ribbon as seen below.

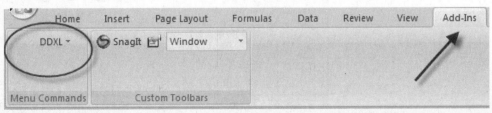

DETERMINING A CONFIDENCE INTERVAL FOR A POPULATION PROPORTION:

To determine the confidence interval for a population proportion using **DDXL** we will begin by considering the Global Warming discussion found in the Chapter Problem at the start of Chapter 7. You will need to determine the number of successes in this problem. Since 70% of the 1501 people polled believe in global warming we can determine that the number of successes is 1051.

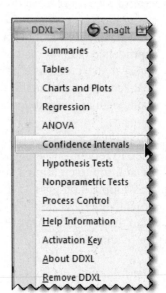

1) Enter the number of successes in cell A1 and the number of trials in cell B2.

2) **Highlight** columns A and B. This is an important step so don't forget to do this. If you forget to highlight the columns which contain the number of successes and trials you will not be able to determine the confidence interval using DDXL.

3) Click on the **Add-ins tab.**

4) Click on the **DDXL icon** in the **Menu commands** group on the left side of the **Add-ins** ribbon. Scroll down the list of options to **Confidence Intervals** and click to select this option.

5) Select **Summ 1 Var Prop Interval** from the Confidence Intervals Dialog box that opens.

6) **Select the column that contains the number of successes** (in this case column A) and **drag it** from the Names and Columns field to the Num Successes field. Next **select the column that contains the number of trials** (in this case column B) and **drag it** from the Names and Columns field to the Num Trials field.

Screen shots for steps (5) and (6) can be seen here:

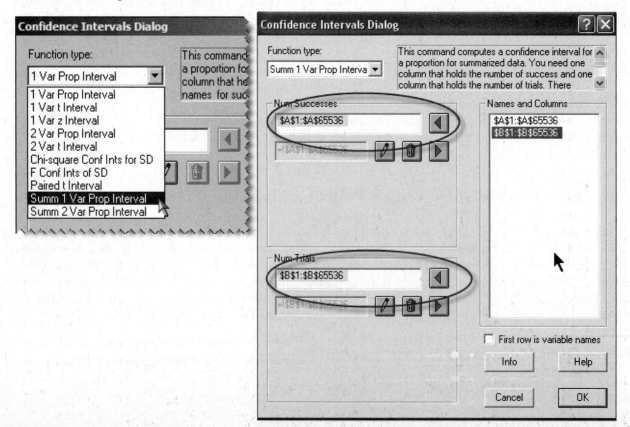

7) Once you have successfully entered the cells into the appropriate place click **OK.**

In the Proportion Interval Setup dialog box

    a. Select the appropriate level of confidence and click on it. In this case we are using **95%.**

    b. Click on **Compute Interval**.

A summary dialog box similar to the one shown here displays the **confidence interval** for the population proportion. Compare these results with those found in your textbook.

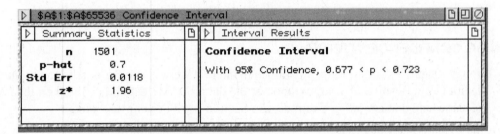

It is possible to **copy and paste your DDXL results** to your Excel worksheet.

1) After you get your results in DDXL simply click on the title bar of the window you wish to copy into Excel. In this case we would want to copy the Confidence Interval window.

2) From the **Edit** menu choose **Copy Window**.

3) Switch to Excel and choose **Paste.**

TO PRACTICE THESE SKILLS

You can practice the technology skills learned in this section by working through the following problems found in your textbook.

1) Use **DDXL** to work through problems 21 and 23 in Section 7-2 Basic Skills and Concepts. Note that the problem asks for the confidence interval as a percentage. DDXL will return a decimal value. You will need to rewrite your results in the appropriate format.

2) Use **DDXL** to work through problem 31 in Section 7-2 Basic Skills and Concepts. Note that you will have to determine the number of successes before you begin.

3) Use **DDXL** to work through problem 35 in Section 7-2 Basic Skills and Concepts.

4) Use the data found in **Data Set 18** in appendix B of your textbook or the data file "M&M" found on the CD data disk to work through problem 45 in Section 7-2 Basic Skills and Concepts.

SECTION 7-3: ESTIMATING A POPULATION MEAN: σ KNOWN

In Section 7-2 we used a sample proportion as the best point estimate of the population proportion. In this section we will apply some of those same ideas to the sample mean $\overline{X}$ as the best point of the population mean μ. Remember that **confidence intervals** allow us to gauge how accurately a sample parameter approximates the same parameter with respect to the population. It gives us information that enables us to better understand the accuracy of the point estimate of a population parameter. The formula for the confidence interval for the population mean when the standard deviation σ is known is given by

$$\overline{X} - E < \mu < \overline{X} + E \ \text{ where } E = z_{\alpha/2} \cdot \frac{\sigma}{\sqrt{n}}$$

To determine a confidence interval for the mean in Excel you will need to know the value for the sample mean, $\overline{X}$, and the value for the population standard deviation, σ. It is relatively easy to find both of these values using the built in statistical functions in Excel or from Descriptive Statistics found in the Data Analysis Tools.

DETERMINING CONFIDENCE INTERVALS

As we work through the process of building a confidence interval we will use the data found in **Data Set 1** in Appendix B (Health Exam Results) of your textbook or the data file "MHealth.xls" found on the CD data disk. This is the same data that is used in Example 1 in Section 7 – 3 of your textbook.

- Open the Excel file MHealth.xls

- **Copy column D** titled **WT** (for Weight) into a new worksheet.

To find the **confidence interval** we make use of the **Insert Function** feature introduced in previous chapters.

1) Locate the **Insert Function** icon on the left side of the **Formula** ribbon in the **Function Library** group.

2) We begin by using the **Insert Function icon** to **determine the sample mean and standard deviation** for our data. If done correctly you should find that the mean for weights is 172.6 lb and the standard deviation is 26. Your decimal values may vary slightly depending on the formatting of the cells you are using.

3) Use the **Insert Function icon** again to search for the function that will allow us to determine the **confidence interval.**

4) In the **Insert Function** dialog box select the **Statistical** function.

5) Scroll through the list of Function names until you see **CONFIDENCE**. Click on this function name.

6) Click **OK**.

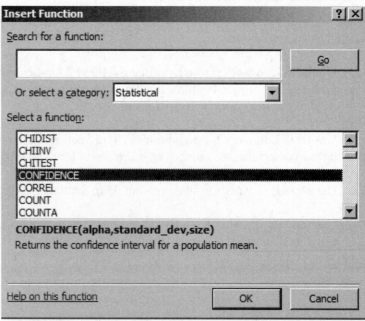

7) This opens a Confidence dialogue box. Assuming a degree of confidence of 95% we can fill in the information required.

    a.   With a degree of confidence of 95%, α = **0.05.**

    b.   Enter the standard deviation that you found in step (2) and the size of the sample, in this case 40.

8) Your dialog box should look similar to the screenshot seen here.

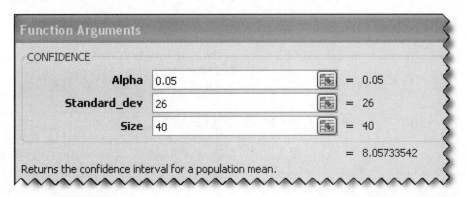

The value returned is called the **margin of error** (or **maximum error**) and is denoted by the letter *E* in the formula presented at the beginning of this section. You now have all of the information you need to create a confidence interval.

1) Determine the upper confidence interval limit ($\overline{X} + E$) and the lower confidence interval limit ($\overline{X} - E$).

	C	D
	mean	172.6
	st dev	26
	margin of error	8.057335
	upper class limit	180.6
	lower class limit	164.5

    a.   The lower limit $\overline{X} - E$ is found using the formula = (D1-D4)

    b.   The upper limit $\overline{X} + E$ is found by using the formula = (D1 + D4)

2) This will give you the following confidence interval 164.5 < μ <180.6. Compare this result with the result found in part (b) of the worked out solution to Example 1 in Section 7 – 3 of your textbook.

CONFIDENCE INTERVALS WITH DDXL

You may also determine a confidence interval for a population mean when σ is known by using the DDXL add-in. You are strongly encouraged to try both methods (using the built-in functions of Excel and the DDXL add-in) in order to determine the advantages and disadvantages of each method.

To use DDXL to determine a confidence interval it will be necessary for you to enter your data in a single column. We will use the same data set that we used above. If you saved that file after finishing the example above you will need to open it now. If you did not save that file you will need to open the Excel file MHealth.xls and to copy column D titled WT (for Weight) into a new worksheet as you did before.

To find the **confidence interval** using DDXL:

1) Once you have copied the WT column to a new worksheet you will need to determine the sample mean and standard deviation as you did before.

2) Click on the **Add-ins tab.**

3) Click on the **DDXL icon** in the **Menu commands** group on the left side of the **Add-ins** ribbon. Scroll down the list of options to **Confidence Intervals** and click to select this option.

4) Select **1 Var z Interval**.

5) Click on the **pencil icon** and enter the range of data. In this problem the data range is A2:A41.

6) Click **OK.**

7) In the dialog box that looks like the one seen on the right

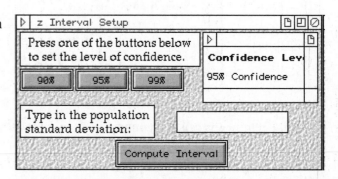

    a. Select the appropriate level of confidence, in this case **95%**.

    b. Enter the standard deviation for the sample.

    c. Click on **Compute Interval**.

A summary dialog box similar to the one shown on the right displays the confidence interval for the population proportion. Compare this to the result we determined previously in Excel by finding the margin of error $(164.5 < \mu < 180.6)$.

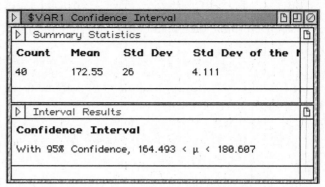

As with our previous DDXL problem, it is possible to copy and paste your results to your Excel worksheet.

TO PRACTICE THESE SKILLS

The following problems will help you to practice the technology skills introduced in this section. When working with data we recommend that you use both Excel and DDXL in order to become familiar with both approaches.

1) Use Excel to work through problem 21 in Section 7-3 Basic Skills and Concepts.

2) Use Excel and DDXL to work through problem 29 in Section 7-3 Basic Skills and Concepts.

3) Use Excel and DDXL to work through problem 30 in Section 7-3 Basic Skills and Concepts.

SECTION 7-4: ESTIMATING A POPULATION MEAN: σ NOT KNOWN

In this section we will present a method for determining a confidence interval estimate for the population mean, μ without the requirement that σ be known. The student t distribution is used to estimate the population mean. The sample confidence interval is $\overline{X} - E < \mu < \overline{X} + E$ where $E = t_{\alpha/2} \cdot \dfrac{s}{\sqrt{n}}$.

Note that the sample standard deviation s, replaces the population standard deviation σ in the formula. To determine the small sample confidence intervals or the population mean with Excel, use the **TINV** function to determine the appropriate t **values**.

In this section we will work with the information presented in Example 2 of Section 7 – 4 in your textbook. This example is titled Constructing a Confidence Interval: Garlic for Reducing Cholesterol. Using an example found in your book will allow us to compare our Excel result with the result presented by the author of the textbook.

Since we are using an example from the textbook rather than actual data we will not need to determine the mean or standard deviation as we did in the previous section.

We need the following information from Example 2:
- the **sample size is 49**
- the **mean is 0.4**
- the **standard deviation is 21.0**

We will use these sample statistics to construct a 95% confidence interval estimate.

DETERMINING CONFIDENCE INTERVAL USING A T DISTRIBUTION

To determine the t **value** we will follow the same method used to find the confidence interval outline in Section 7–3. We will need to determine the margin of error E which can be determined easily using the built in function in Excel.

To find the TINV **confidence interval** we make use of the **Insert Function** feature introduced in previous chapters.

1) Locate the **Insert Function** icon on the left side of the **Formula** ribbon in the **Function Library** group.

2) In the **Insert Function** dialog box select the **Statistical** function.

3) Scroll through the list of Function names until you see **TINV.** Click on this function name.

4) Click **OK**.

The following dialogue box will open in your Excel workbook.

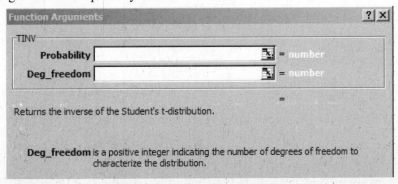

Assuming a degree of confidence of 95% we can fill in the information required. Recall that with a degree of confidence of 95%, the probability = 0.05, the deg_freedom is 48.

1) This returns a **t-distribution** of 2.011.

2) **To find our confidence interval** complete the information as seen on the Excel worksheet on the right.

K	L
mean	0.4
st dev	21
sample size	49
t distribution	2.011
error (E)	6.033
lower limit	-5.6
upper limit	6.4

   a. **To determine the value of E** use the formula $E = t_{\alpha/2} \cdot \dfrac{s}{\sqrt{n}}$.

   b. Use $\overline{X} - E < \mu < \overline{X} + E$ to find the upper and lower limit of the confidence interval.

3) This will give you the following confidence interval $-5.6 < \mu < 6.4$. Compare this result with the result found in the worked out solution to Example 2 in Section 7 – 4 of your textbook.

CONFIDENCE INTERVALS WITH DDXL

You may also determine a confidence interval for a population mean when σ is not known using DDXL. You are strongly encouraged to try both methods (Excel functions and the DDXL add-in) in order to determine the advantages and disadvantages of each method.

We cannot use DDXL for the example we used above. We need to have a data set with which to determine our confidence interval.

To use DDXL to determine a confidence interval when σ is not known simply follow the instructions outlined in the previous section with one change – choose the **1 Var t Interval.**

TO PRACTICE THESE SKILLS

The following problems will help you to practice the technology skills introduced in Section 7-4. When working with data we recommend that you use both Excel and DDXL in order to become familiar with both approaches.

1) Use Excel to work through problem 13 in Section 7-4 Basic Skills and Concepts.

2) Use Excel to work through problem 17 in Section 7-4 Basic Skills and Concepts.

3) Use Excel and DDXL to work through problem 27 in Section 7-4 Basic Skills and Concepts.

4) Use Excel and DDXL to work through problem 31 in Section 7-4 Basic Skills and Concepts.

CHAPTER 8: HYPOTHESIS TESTING

SECTION 8-1: OVERVIEW

Setting up and testing a hypotheses is an important part of statistical inferences To In order to construct a hypothesis test we start with a conclusion that has been presented, either because it is believed to be true or because it is to be used as a basis for an argument. Generally this hypothesis has not been proven, as in the chapter opener regarding an effective method for gender selection.

In this chapter we will look at the statistical process that is used for testing a claim made about a population. Recall that in Chapter 7 we used sample statistics to estimate population parameters; in this chapter we will use sample statistics to test hypotheses (or claims) that have been made about population parameters. While Excel has many built in statistical analysis tools available, it does not have a built in tool or function that will allow us to directly perform hypothesis tests. We will once again make use of the DDXL add-in as we work through the various hypotheses tests we will perform in this chapter.

SECTION 8-2: TESTING A CLAIM ABOUT A PROPORTION

It is important that you spend a good amount of time in Section 8 – 2 of your textbook in order to become comfortable with the different individual components of a hypothesis test. Once you are comfortable with this material you will be ready to use those components to test claims made about population proportions (which coincides with Section 8 – 3 of your textbook). We will make use of the DDXL add-in to perform a z test of the hypothesis for a proportion.

Z TEST FOR ONE VARIABLE PROPORTION TEST:

The Chapter Problem found at the beginning of Chapter 8 focused on the use of a newer gender selection method called MicroSort. The MicroSort XSORT method is designed to increase the likelihood of having a baby girl while the YSORT method is designed to increase the likelihood of having a baby boy. In the Chapter opener it was noted that "Among 726 babies born to couples using the XSORT method in an attempt to have a baby girl, 668 of the babies were girls and the others were boys." We will test the claim that "with the XSORT method the proportion of girls is greater than 0.5". This problem is outlined in the **Testing the Effectiveness of the MicroSort Method of Gender Selection** example found in Section 8-3 of your textbook.

1) Enter the number of successes in cell A1 and the number of trials in cell B2.

2) **Highlight** columns A and B. This is an important step so don't forget to do this. If you forget to highlight the columns which contain the number of successes and trials you will not be able to perform a hypothesis test using DDXL.

3) Click on the **Add-ins tab.**

4) Click on the **DDXL icon** in the **Menu commands** group on the left side of the **Add-ins** ribbon. Scroll down the list of options to **Hypothesis Tests** and click to select this option.

5) When the **Hypothesis Tests** dialog box opens select **Summ 1 Var Prop Test** from Function type.

6) **Select the column that contains the number of successes** (in this case column A) and **drag it** from the Names and Columns field to the Num Successes field. Next **select the column that contains the number of trials** (in this case column B) and **drag it** from the Names and Columns field to the Num Trials field.

7) Click **OK**.

8) This will open a dialog box that will require additional information. Complete each of the four steps in the dialog box as outlined on the following page.

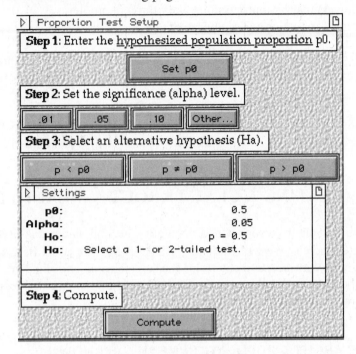

a. **Step 1:** Click on **"Set p0 "**. We will enter the hypothesized test proportion found within the example ($p = 0.5$). Enter the value in the appropriate spot and click **OK.**

b. **Step 2:** Set the **significance level** by clicking on the appropriate value. In this case use a significance level of 0.05.

c. **Step 3**: Select the alternative hypothesis. In this problem the null hypothesis states H_0: $p = 0.5$. Therefore the alternative hypothesis is H_1: $p > 0.5$. Choose p > p0.

d. **Step 4**: Click on **Compute.**

9) **DDXL** presents the following results that include the test statistic, the P – value and a conclusion to reject the null hypothesis.

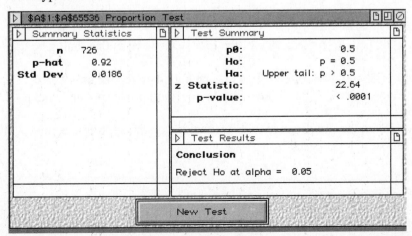

TO PRACTICE THESE SKILLS

You can apply the technology skills learned in this section by working on the following problems. In each problem you will need to use the DDXL add-in.

1) To practice testing claims about proportions work through problems 9, 11, 15 and 21 found in the Basic Skills and Concepts for Section 8-3 of your textbook.

2) Use Data Set 18 to work through problem 33 in the Basic Skills and Concepts for Section 8-3 of your textbook. This problem gives you an opportunity to start with actual data to test a claim about proportions.

3) Use Data Set 6 to work through problem 35 in the Basic Skills and Concepts for Section 8-3 of your textbook. This problem gives you an opportunity to start with actual data to test a claim about proportions.

SECTION 8-3: TESTING A CLAIM ABOUT A MEAN: σ KNOWN

In this section we will test claims made about a population mean μ, and we assume that the population standard deviation σ is known. We will work with Data Set 1 in Appendix B. Copy the column titled Weight (WT) into a new worksheet and determine the sample mean and standard deviation. Recall that these functions can be accessed by choosing the **Insert Function** icon located on the left side of the **Formula** ribbon in the **Function Library** group.

TESTING CLAIMS ABOUT A POPULATION MEAN μ (σ KNOWN)

This problem is outlined in the *P*-Value Method example found in Section 8-4 of your textbook. We will test the claim that men have a mean weight greater than 166.3 lb.

1) Click on the **Add-ins tab.**

2) Click on the **DDXL icon** in the **Menu commands** group on the left side of the **Add-ins** ribbon. Scroll down the list of options to **Hypothesis Tests** and click to select this option.

3) When the **Hypothesis Tests** dialog box opens select **1 Var z Test** from Function type.

4) Click on the pencil icon and list the range of cells that include your data. (In this example your range of cells should be A2:A41).

5) Click **OK**.

6) A dialog box will open. Complete each of the four steps listed in that dialog box as follows:

> **Step 1:** Click on **"Set μθ and sd"** . The hypothesized mean is 172.6. The standard deviation is 26.3. Enter these values in the appropriate spot and click **OK.**

> **Step 2:** Set the **significance level** by clicking on the appropriate value. In this case use a significance level of 0.05.

Step 3: Select the **alternative hypothesis**. In this problem the null hypothesis states that the mean weight is 166.3. The alternative hypothesis states that the mean weight is greater than 166.3. Choose $\mu > \mu\theta$.

Step 4: Click on **Compute**.

7) DDXL presents the following results that include the test statistic, the P – value and a conclusion to fail to reject the null hypothesis.

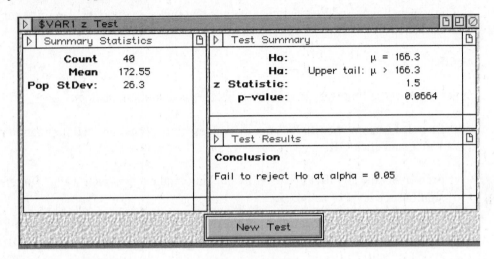

TO PRACTICE THESE SKILLS

You can practice the skills learned in this section by working through the following problems. Be sure to use the DDXL add-in for each of these problems.

1) Use Data Set 18 found in Appendix B or in the data file M&M.xls to work through problem 9 in the Basic Skills and Concepts for Section 8-4 of your textbook.

2) Use Data Set 22 found in Appendix B or in the data file GARBAGE.xls to work through problem 18 in the Basic Skills and Concepts for Section 8-4 of your textbook.

SECTION 8-4: TESTING A CLAIM ABOUT A MEAN: σ NOT KNOWN

In this section we turn our attention to testing claims made about a population mean μ when the population standard deviation σ is not known. We will use the Student t distribution rather than the normal distribution used in the preceding section where σ was known. We will rely on the DDXL add-in to perform a t test of the hypothesis of the mean. This test compares the observed t test statistic to the point of the t distribution that corresponds to the test's chosen α level.

TESTING CLAIMS ABOUT A POPULATION MEAN μ (WITH σ NOT KNOWN):

We will work with Data Set 1 in Appendix B. Copy the column titled Weight (WT) into a new worksheet.

This problem is outlined in Example 1 from Section 8 – 5 of your textbook. We will not assume that the value of the population mean, σ, is known. We will test the claim that men have a mean weight greater than 166.3 lb.

These directions should begin to look familiar to you

1) Click on the **Add-ins tab.**

2) Click on the **DDXL icon** in the **Menu commands** group on the left side of the **Add-ins** ribbon. Scroll down the list of options to **Hypothesis Tests** and click to select this option.

3) When the **Hypothesis Tests** dialog box opens select **1 Var t Test** from Function type.

4) Click on the pencil icon and list the range of cells that include your data. (In this example your range of cells should be A2:A41).

5) Click **OK**.

6) Complete each of the four steps listed in the dialog box as outlined below.

> **Step 1:** Click on **"Set $\mu\theta$"**. Enter the values for the hypothesized population mean (H_o: $\mu =$ 166.3) and click **OK**.
>
> **Step 2:** Set the **significance level** by clicking on the appropriate value. In this case use a significance level of $\alpha = 0.05$.
>
> **Step 3:** Select the **alternative hypothesis**. In this problem the alternative hypothesis states that we test the claim that men have a mean weight greater than 166.3 lb. Therefore we choose $\mu > \mu\theta$.
>
> **Step 4:** Click on **Compute.**

7) **DDXL** presents the following results that include the test statistic, the P – value and a conclusion to fail to reject the null hypothesis.

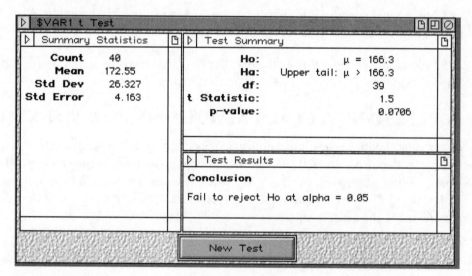

TO PRACTICE THESE SKILLS

You can practice the technology skills learned in this section by working on the following problems using the DDXL add-in.

1) Work through problem 17 in the Basic Skills and Concepts for Section 8-5 of your textbook.

2) Use Data Set 16 found in Appendix B or in the data file CARS.xls to work through problem 26 in the Basic Skills and Concepts for Section 8-5 of your textbook.

3) Use Data Set 2 found in Appendix B or in the data file BODYTEMP.xls to work through problem 31 in the Basic Skills and Concepts for Section 8-5 of your textbook.

CHAPTER 9: INFERENCES FROM TWO SAMPLES

SECTION 9-1: OVERVIEW

In Chapters 7 and 8 we used sample data to construct confidence interval estimates of population parameters and to test hypotheses about certain population parameters. In each case we used one single sample to form an inference about single population. In this chapter we will extend the methods we have learned to include confidence intervals and hypothesis tests for comparing two sets of sample data. We will utilize the capabilities of Excel as well as the DDXL add-in to test the hypothesis made about two population means.

- the Data Analysis **z-Test: Two Samples for Means.**

- the DDXL add-in Hypothesis Tests: **Summ 2 Var Prop Test**

    Confidence Intervals : **Summ 2 Var Prop Test**

t-Test: Paired Samples for Means
This analysis tool and its formula perform a paired two-sample student's t-test to determine whether a sample's means are distinct. This t-test form does not assume that the variances of both populations are equal. You can use a paired test when there is a natural pairing of observations in the samples, such as when a sample group is tested twice — before and after an experiment.

F Test Two Samples for Variances
This analysis tool performs a two-sample F-test to compare two population variances and to determine whether the two population variances are equal. It returns the p value of the one tailed F statistic, based on the hypothesis that array 1 and array 2 have the same variance.

t-Test: Two Samples Assuming Equal Variances
This test calculates a two sample Student t Test. The test assumes that the variance in each of the two groups is equal. The output includes both one tailed and two tailed critical values.

t-Test: Two Samples Assuming Unequal Variances.
This test calculates a two sample Student t Test. The test allows the variances in the two groups to be unequal. The output includes both one tailed and two tailed critical values.

SECTION 9-2: INFERENCES ABOUT TWO PROPORTIONS

When using sample data to compare two population proportions we will use DDXL and the **Summ 2 Var Prop Test.**   Since we are comparing two population proportions we will need to enter the number of successes as well as the number of trials for both Sample 1 and Sample 2 into Excel.

Using the information presented in Example 1 in Section 9 – 2 of your textbook titled **Do Airbags Save Lives?** In this example the two samples we are dealing with are those with airbags and those with no airbags. Use a 0.05 significance level to test the claim that the fatality rate of occupants is lower for those in cars equipped with airbags.

DDXL – SUMM 2 VAR PROP TEST

1) Enter the number of successes for the first sample in cell A1. In cell B1 enter the number of trials for the first sample. In cell C1 enter the number of successes for the second sample. In cell D1 enter the number of trials for the second sample.

2) **Highlight** columns A, B, C and D. This is an important step so don't forget to do this. If you forget to highlight the columns which contain the number of successes and trials for both samples you will not be able to perform a hypothesis test using DDXL.

3) Click on the **Add-ins tab.**

4) Click on the **DDXL icon** in the **Menu commands** group on the left side of the **Add-ins** ribbon. Scroll down the list of options to **Hypothesis Tests** and click to select this option.

5) When the **Hypothesis Tests** dialog box opens select **Summ 2 Var Prop Test** from Function type.

6) In the Hypothesis dialog box **select the column that contains the number of successes for the first sample** (in this case column A) and **drag it** from the Names and Columns field to the Num Successes 1field. Next **select the column that contains the number of trials for the first sample** (in this case column B) and **drag it** from the Names and Columns field to the Num Trials 1 field. Repeat this process for Num Success 2 field and the Num Trials 2 field.

7) Click **OK**.

8) Follow these steps in the **Proportion Test Setup** dialog box:

   a. **Step 1:** This step is optional so you can skip it or set the difference at 0 (since we will only be testing claims that p1 = p2).

   b. **Step 2**: For this example set the significance level at 0.05.

   c. **Step 3**: Select the appropriate alternative hypothesis, in this case **p1 – p2 < p**.

   d. **Step 4**: Click on **Compute**.

9) DDXL returns the value for test statistic, the p –value as well as the conclusion to reject the null hypothesis. Compare these results with those found in your textbook.

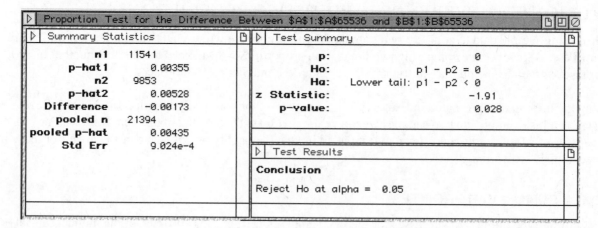

DDXL - CONFIDENCE INTERVALS FOR PROPORTION PAIRS

We can construct a confidence interval estimate of the difference between population proportions for the **Do Airbags** example that we just looked at.   This corresponds to Example 2 in Section 9 – 2 of your textbook.

1) If you did not enter the information used in the previous section or did not save the information you will begin by entering the number of successes for the first sample in cell A1. In cell B1 enter the number of trials for the first sample. In cell C1 enter the number of successes for the second sample. In cell D1 enter the number of trials for the second sample.

2) **Highlight** columns A, B, C and D. This is an important step so don't forget to do this. If you forget to highlight the columns which contain the number of successes and trials for both samples you will not be able to perform a hypothesis test using DDXL.

3) Click on the **Add-ins tab.**

4) Click on the **DDXL icon** in the **Menu commands** group on the left side of the **Add-ins** ribbon. Scroll down the list of options to **Confidence Interval** and click to select this option.

5) When the **Confidence Interval** dialog box opens select **Summ 2 Var Prop Interval** from Function type.

6) In the Confidence Intervals dialog box **select the column that contains the number of successes for the first sample** (in this case column A) and **drag it** from the Names and Columns field to the Num Successes 1field. Next **select the column that contains the number of trials for the first sample** (in this case column B) and **drag it** from the Names and Columns field to the Num Trials 1 field.   Repeat this process for Num Success 2 field and the Num Trials 2 field.

7) Click **OK**.

8) A dialog box will open in order for you to set the appropriate confidence level. In this case we choose 90%.

9) Click on **Compute Interval.**

10) The following results are displayed.

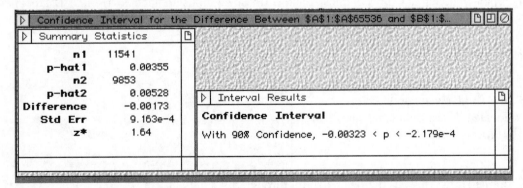

Compare these results with those found in your textbook. You may notice that the value on the right side of the interval in the DDXL result appears to be different than the value found in your textbook. DDXL has returned a value of $-2.19e^{-4}$ which is similar to the scientific notation you would see on a graphing calculator. Note that $-2.19e^{-4}$ is equivalent to -0.000219 which is similar to the answer found in your textbook.

TO PRACTICE THESE SKILLS

You can practice the technology skills covered in this section by working on the following problems found in your textbook.

1) Work on problems 7, 11, and 13 in the Basic Skills and Concepts for Section 9-2 of your textbook. These problems focus on hypothesis testing and clearly identify the number of successes and trials for each sample.

2) Work on problems 9, 15 and 21 in the Basic Skills and Concepts for Section 9-2 of your textbook. These problems focus on determining confidence interval estimates.

SECTION 9-3: INFERENCES ABOUT TWO MEANS: INDEPENDENT SAMPLES

In this section we will work with sample data from two independent samples to test the hypothesis made about two population means. We will also construct confidence interval estimates of the differences between two population means. We will outline the following methods:

- the Data Analysis **t-Test: Two-Sample Assuming Unequal Variances**

- the DDXL add-in Hypothesis Tests: **2 Var t Test**

Excel requires that you use original lists of sample data. Example 4 **Are Men and Women Equal Talkers** in Section 9 -3 uses all of the values in Data Set 8. We will work through an example similar to Example 4 but we will only use the data found in columns A and B. We will use a 0.05 significance level to test the claim that men and women speak the same mean number of words in a day.

T-TEST: TWO-SAMPLE ASSUMING UNEQUAL VARIANCES:

1) Open Data Set 8. Copy columns A and B (these columns are labeled 1M and 1F) to a new worksheet.

2) Click on the **Data** tab.

3) Click on the **Data Analysis** icon in the **Analysis group** which is located on the right hand side of the **Data** ribbon.

4) In the **Data Analysis** dialog box scroll through the **Analysis Tools** list to select **t-Test: Two-Sample Assuming Unequal Variances**. You can see this on the following page.

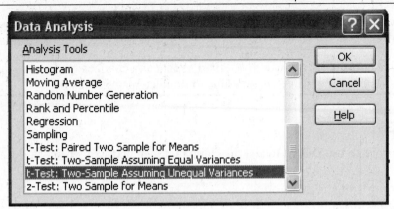

5) Click on **OK.**

6) In the **t-Test: Two-Sample Assuming Unequal Variances** dialog box enter the following
    information:

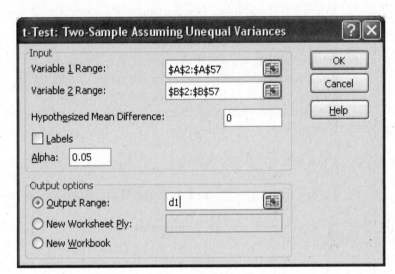

    a. The range of cells
       containing the number of
       words in the column
       labeled 1M in the
       Variable 1 Range box.

    b. The range of cells
       containing the number of
       words in the column
       labeled 1F in the
       Variable 2 Range box.

    c. Enter 0 in the
       **Hypothesized Mean
       Difference** box or just
       leave it blank.

    d. Enter 0.05 in the **Alpha** box.

    e. Determine where you wish to display the output.

    f. Click **OK**.

7) The following summary of
    information containing calculations
    for the t-test is added to your current
    Excel worksheet.

t-Test: Two-Sample Assuming Unequal Variances		
	Variable 1	Variable 2
Mean	16576.11	18443.1986
Variance	61960358	55645786.51
Observations	56	56
Hypothesized Mean Difference	0	
df	110	
t Stat	-1.28838	
P(T<=t) one-tail	0.100158	
t Critical one-tail	1.658824	
P(T<=t) two-tail	0.200316	
t Critical two-tail	1.981765	

Based on the results we fail to reject the null hypothesis.

Note:

It is worth mentioning that this chart is not a "live" chart so that any changes made to the original data at this point would require using Data Analysis a second time to produce new test results.

DDXL – 2 VAR T TEST

Use the same data as above to use DDXL to test the claim that men and women speak the same mean number of words in a day.

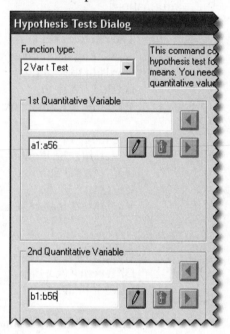

1) Click on the **Add-ins tab.**

2) Click on the **DDXL icon** in the **Menu commands** group on the left side of the **Add-ins** ribbon. Scroll down the list of options to **Hypothesis Tests** and click to select this option.

3) When the **Hypothesis Tests** dialog box opens select **2 Var t Test** from Function type.

4) In the dialog box click on the pencil icon for the **1st Quantitative Variable** and enter the range of data for columns A. Then click on the pencil icon for the **2nd Quantitative Variable** and enter the range of data for column B.

5) Click **OK**

6) Follow these steps in the **2 Sample t Test Setup** dialog box:

   a. **Step 1**: Select **2-sample.**

   b. **Step 2**: This step is optional so you can skip over it or set the difference at 0.

   c. **Step 3**: Set the significance level at 0.05.

   d. **Step 4**: Select the appropriate alternative hypothesis.

   e. **Step 5**: Click on **Compute**.

7) The results can be seen on the following page. In addition to the Test Summary, DDXL also returns the mean and standard deviation of each variable and a conclusion to fail to reject the null hypothesis.

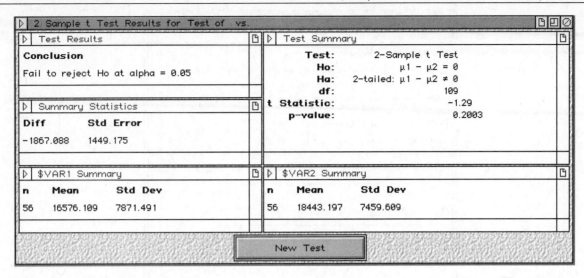

CONFIDENCE INTERVAL ESTIMATES

We continue to work with the same sample data that we used in the two methods preceding this section on confidence intervals. As a reminder, we used a 0.05 significance level to test the claim that men and women speak the same mean number of words in a day.

Switching our focus from hypothesis tests to confidence intervals, we can construct a 95% confidence interval estimate of the difference between the mean number of words spoken by men and the mean number of words spoken by women. This will be done using the DDXL add-in. The procedure for doing this is very similar to the one used to determine the 2 Var t Test.

1) Click on the **Add-ins tab.**

2) Click on the **DDXL icon** in the **Menu commands** group on the left side of the **Add-ins** ribbon. Scroll down the list of options to **Confidence Intervals** and click to select this option.

3) When the **Confidence Interval** dialog box opens select **2 Var t Interval** from Function type.

4) In the dialog box click on the pencil icon for the **1st Quantitative Variable** and enter the range of data for columns A. Then click on the pencil icon for the **2nd Quantitative Variable** and enter the range of data for column B.

5) Click **OK**

6) In **2 Sample t Interval Setup**

    a.   **Step 1:** Choose **2 sample**.

    b.   **Step 2**: Select the appropriate confidence level (in this case 95%)

    c.   **Step 3:** Click on **Compute Interval**.

7) The results are displayed on the next page. In addition to the Confidence Interval results, DDXL also returns the mean and standard deviation for each variable.

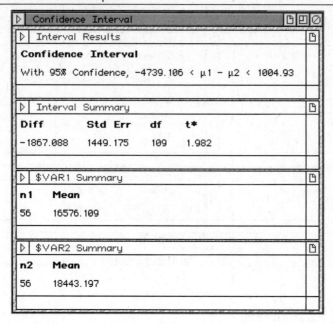

TO PRACTICE THESE SKILLS

You can practice the skills learned in this section by working through the following problems found in your textbook.

1) Work on problems 29 and 30 in the Basic Skills and Concepts for Section 9-3 of your textbook.

2) Use Data Set 9 to work through problem 33 in the Basic Skills and Concepts for Section 9-3 of your textbook.

SECTION 9-4: INFERENCES FROM DEPENDENT SAMPLES

In the previous section we worked with independent populations. We now turn our attention to confidence interval estimates and hypothesis tests for dependent samples or matched pairs. The analysis can be done using either

- the Data Analysis **t-Test: Paired Two Samples for Means.**

- the DDXL add-in Hypothesis Tests: **Paired t Test**

Begin by entering the data found in Table 9–1 from the example **Hypothesis Test of Claimed Freshman Weight Gain** into an Excel spreadsheet. This example is found in Section 9–4 of your textbook. The data needs to be entered in columns rather than rows with the April weight in column A and the September weight in column B. In column C show the differences between each April weight and September weight as shown in your textbook. The Data Analysis t Test requires that the data for each group be in a separate column. This is often referred to as **unstacked data.**

T-TEST: PAIRED TWO SAMPLES FOR MEANS

Using the data entered into Excel with a 0.05 significance level we will test the claim that for the population of students, the mean change in weight from September to April is equal to 0 kg.

1) Click on the **Data** tab.

2) Click on the **Data Analysis** icon in the **Analysis group** which is located on the right hand side of the **Data** ribbon.

3) In the **Data Analysis** dialog box scroll through the **Analysis Tools** list to select **t-Test: Paired Two Samples for Means**.

4) Click **OK**.

5) In the t Test: Paired Two Samples for Means dialog box enter the following information

   a. The cell range containing April weights in the **Variable 1 Range** box.

   b. The cell range for September weights in the **Variable 2 Range** box.

   c. Enter 0.05 in the **Alpha** box. This value is supplied in the problem.

   d. Determine where you wish to display the output.

   e. Click **OK**.

t-Test: Paired Two Sample for Means

Input	
Variable 1 Range:	$A$2:$A$6
Variable 2 Range:	$B$2:$B$6
Hypothesized Mean Difference:	
☐ Labels	
Alpha:	0.05

Output options	
⦿ Output Range:	D1
○ New Worksheet Ply:	
○ New Workbook	

6) The following summary of information containing calculations for the t-test is in added to your current Excel worksheet.

t-Test: Paired Two Sample for Means		
	Variable 1	Variable 2
Mean	65.2	65
Variance	57.7	52.5
Observations	5	5
Pearson Correlation	0.949333347	
Hypothesized Mean Difference	0	
df	4	
t Stat	0.187317162	
P(T<=t) one-tail	0.430264858	
t Critical one-tail	2.131846782	
P(T<=t) two-tail	0.860529716	
t Critical two-tail	2.776445105	

In addition to the test statistic, and the corresponding critical values (circled above) Excel also displays the P values for a one and two tailed test. Compare the test statistic and the critical values found with Excel with those found in Step 6 of the solution for the example **Hypothesis Test of Claimed Freshman Weight Gain** in your textbook.

You can also use DDXL to perform the hypothesis test for this problem.

DDXL – PAIRED T TEST

1) Click on the **Add-ins tab.**

2) Click on the **DDXL icon** in the **Menu commands** group on the left side of the **Add-ins** ribbon. Scroll down the list of options to **Hypothesis Tests** and click to select this option.

3) When the **Hypothesis Tests** dialog box opens select **Pair t Test** from Function type.

4) In the dialog box click on the pencil icon for the **1st Quantitative Variable** and enter the range of data for columns A. Then click on the pencil icon for the **2nd Quantitative Variable** and enter the range of data for column B.   Click on the pencil icon for the **Pair Labels** and enter the range of data for the difference between the April weight and the September weight.

5) Click **OK**

6) Complete these steps in the **paired t test** dialog box.

   a. **Step 1**: This step is optional so you can skip it.

   b. **Step 2**: Set the significance level. In this problem it is 0.05.

   c. **Step 3**: Select the appropriate alternative hypothesis.

   d. **Step 4**: Click on **Compute**.

7) The following results will be displayed. Compare these results with those found in Step 6 of the solution for the example **Hypothesis Test of Claimed Freshman Weight Gain** in your textbook**.**

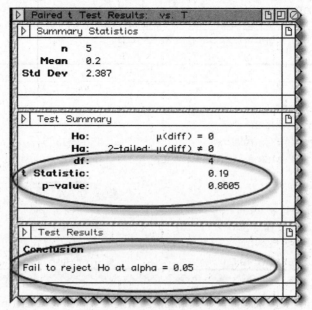

DDXL – CONFIDENCE INTERVALS FOR MATCHED PAIRS

Continue to use the sample data found in Table 9–1 from the example **Hypothesis Test of Claimed Freshman Weight Gain.** This data should already be entered into an Excel spreadsheet. If you did not save your file from the preceding section you will need to enter that information into an Excel worksheet again. The data needs to be entered in columns rather than rows with the April weight in column A and the September weight in column B. In column C show the differences between each April weight and September weight as shown in your textbook.

We will use this sample data to construct a 95% confidence interval estimate of which is the mean of the mean of the "April–September" weight differences of college students in their freshman year. This will be done using the DDXL add-in and the **Paired *t* Interval**. The process for doing this is very similar to the one we used to determine the 2 Var t Test.

1) Click on the **Add-ins tab.**

2) Click on the **DDXL icon** in the **Menu commands** group on the left side of the **Add-ins** ribbon. Scroll down the list of options to **Confidence Intervals** and click to select this option.

3) When the **Confidence Interval** dialog box opens select **Paired *t* Interval** from Function type.

4) I In the dialog box click on the pencil icon for the **1ˢᵗ Quantitative Variable** and enter the range of data for columns A. Then click on the pencil icon for the **2ⁿᵈ Quantitative Variable** and enter the range of data for column B.   Click on the pencil icon for the **Pair Labels** and enter the range of data for the difference between the April weight and the September weight.

5) Click **OK**

6) Select the appropriate confidence level (in this case 95%) and click on **Compute Interval.**

7) The following results are displayed. Compare these results with those found in the solution for Example 3 **Confidence Interval for Estimating the Mean Weight Change** in your textbook.

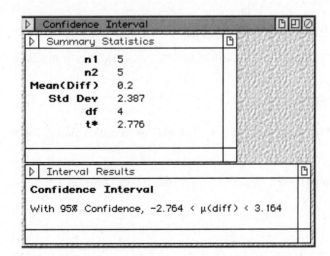

TO PRACTICE THESE SKILLS

Use the data presented in problems 5, 7, 9, 10, 15 and 17 in the Basic Skills and Concepts for Section 9-4 of your textbook to practice the technology skills learned in this section.

- Questions 5 and 7 are paired so that you can test a hypothesis in one problem (#5) and then determine a confidence interval in the second problem (#7) using the same data.

- Questions 9 and 10 are paired so that you can test a hypothesis in one problem (#9) and then determine a confidence interval in the second problem (#10) using the same data.

- Questions 15 and 17 are independent of one another.

SECTION 9-5: COMPARING VARIATION IN TWO SAMPLES

In this section we present a test of hypothesis for comparing two population variances. This will be done using the **F Test Two Sample for Variances** found in Excel's Data Analysis tools.

We will work through Example 1: **Comparing Variations in Weights of Quarters** for this process.

- We need to start by opening the file COINS.xls or by entering the data found in Data Set 20 (Coin Weights) in Appendix B of your textbook.

- Copy the columns for the weights for pre-1964 quarters and post-1964 quarters into a new worksheet.

We will use this information along with the **F Test Two Sample for Variance** found in Excel to work through the **Comparing Variation in Weights of Quarters** example found in Section 9-5 of your statistics book. Use a 0.05 significance level to test the claim that the weights of pre-1964 quarters and the weights of post-1964 quarters are from populations with the same standard deviation.

F-TEST: TWO SAMPLE FOR VARIANCES

If you did not create your data file as outlined at the start of this section please do so before going any further.

1) Click on the **Data** tab.

2) Click on the **Data Analysis** icon in the **Analysis group** which is located on the right hand side of the **Data** ribbon.

3) In the **Data Analysis** dialog box scroll through the **Analysis Tools** list to select **F-Test Two Sample for Variances.**

4) Click **OK**.

5) In the **F- Test: Two Sample for Variances** dialog box enter the following information:

   a. The range of cells containing the sample of weights of pre-1964 quarters in the **Variable 1 Range** box.

b. The range of cells containing the sample of weights of post-1964 quarters in the **Variable 2 Range** box.

c. Enter 0.05 in the **Alpha** box. This value is supplied in the problem.

d. Determine where you wish to display the output.

e. Click **OK**.

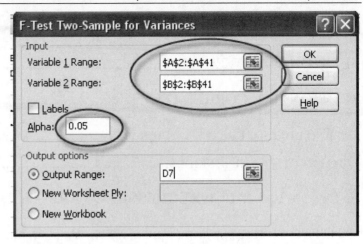

6) The following summary of information containing calculations for the F-test is in added to your current Excel worksheet.

F-Test Two-Sample for Variances		
	Variable 1	Variable 2
Mean	6.1926725	5.6392975
Variance	0.007568136	0.003836163
Observations	40	40
df	39	39
F	1.972839986	
P(F<=f) one-tail	0.018378023	
F Critical one-tail	1.704465067	

Excel will return the F test statistic, the P-value for the one-tailed case, and the critical F value for the one-tailed case. For a two-tailed test you can double the P-value returned by Excel. The test statistic $F = 1.9728$ does fall within the critical region, so we reject the null hypothesis of equal variances. Our conclusion therefore is that there is sufficient evidence to warrant rejection of the claim of equal standard deviations.

TO PRACTICE THESE SKILLS

You can practice the technology skills learned in this section by working on the following problems found in your textbook.

1) Work on problems 15 and 17 in the Basic Skills and Concepts for Section 9-5 of your textbook.

2) Use Data Set 3 found in Appendix B or in the data file FRESH15.xls to work through problem 19 in the Basic Skills and Concepts for Section 9-5 of your textbook.

CHAPTER 10: CORRELATION AND REGRESSION

SECTION 10-1: OVERVIEW

In this chapter we will be working with data that comes in pairs. We will be determining whether there is a relationship between the paired data, and will be trying to identify the relationship if it exists.

Excel provides an excellent tool to help us consider whether there is a statistically significant relationship between two variables. We can create a scatter plot, find a line of regression, and use our regression equation to predict values for one of the variables when we know values of the other variable.

The new functions introduced in this section are outlined below.

CORREL

This returns the correlation coefficient between two data sets. Your paired data must be entered in adjacent columns.

ADD TRENDLINE

This feature adds the linear regression graph to the scatter plot of a set of data values.

REGRESSION

This function returns information on Regression Statistics, as well as other information based on the linear regression equation. Your data values must be entered in adjacent columns.

SECTION 10-2: CORRELATION

In this section, we will take a look at a picture of a collection of paired sample data (duration in seconds of an Old Faithful eruption, interval in minutes after the eruption) to help us determine if there appears to be a relationship between the variable x (duration of an eruption in seconds) and the variable y (interval in minutes after the eruption). We will work with the data from Data Set 15: Old Faithful Geyser in Appendix B in your book.

	A	B
1	**Duration**	**Interval After**
2	240	92
3	237	95
4	250	92
5	243	100
6	255	90
7	120	65
8	260	92
9	178	72
10	259	93
11	245	98

1) **Collect the data you want to work with in Excel:** Open the data set either from the CD with your book, or from the website. In a new worksheet, copy and paste the information on Duration as well as the information on Interval After. Make sure that you keep the columns exactly as they are listed in the original data.

2) **Compute the correlation coefficient:** In cell C1, type" r =", and move your cursor to cell D1. To find the linear correlation coefficient, click on the **Function icon** in the formula bar. In the "Search for a function" box, type in correlation, and then click on **OK.** Click on **CORREL** in the Select a function box**.**
   Then click on **OK**.

   a. In **Array1**, enter the range of cells where the data you want to use on your horizontal axis (x) is stored. In **Array2**, enter the range of cells where the data

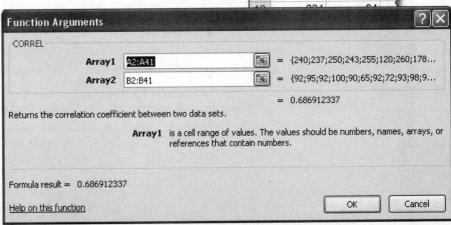

Function Arguments		
CORREL		
Array1	A2:A41	= {240;237;250;243;255;120;260;178...
Array2	B2:B41	= {92;95;92;100;90;65;92;72;93;98;9...
		= 0.686912337

Returns the correlation coefficient between two data sets.

   Array1 is a cell range of values. The values should be numbers, names, arrays, or references that contain numbers.

Formula result = 0.686912337

Help on this function   OK   Cancel you want to use on your vertical axis (y) is stored. Then click on **OK**. You will see the correlation coefficient in cell D1. This value should round to 0.687.

3) **Create the scatter plot:** To create the scatter plot for this data, click on the **Scatter** icon in the **Charts group** in the **Insert** tab, and select the first picture shown. A blank chart will appear in your worksheet.

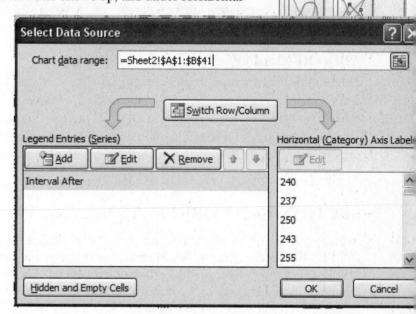

a. **Select the Data you want to graph:** Right click somewhere inside the Chart Area, and choose **Select Data.** The **Select Data Source** box will open. In the **Select Data Source** box, click on the collapse icon for the Chart Data Range box, and select the cells where your paired data is located, including the column names. Notice that underneath the Legend Entries (Series), the word Interval After shows up, and under Horizontal (Category) Axis Labels, the values from the Duration column show up. Click on **OK.**

b. **Modify your initial graph:** You will initially see a graph that looks something like the one shown. You need to make a number of adjustments!

   i. **Delete the legend box:** Clicking on the legend, and then press Delete.

   ii. **Delete the lines connecting the points:** If your graph shows up with lines connecting the points, move your cursor over the cluster of dots, and right click. From the options that appear, choose **Format Data Series.** In the Format Data Series Dialog box, click on **Line Color,** and click in the bubble by **No Line.** Then click **Close.**

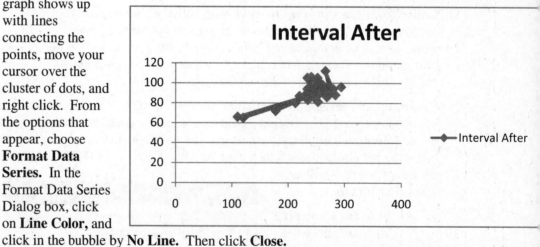

   iii. **Name your graph appropriately:** Click on the name at the top of the graph, and change it to Old Faithful Interval After vs. Duration. If you right click while the Chart Name is selected, you can then make changes in your font size.

iv. **Add axes names:** Click anywhere within the Chart Region. You will see **Chart Tools** appear at the top of your Excel page. Click on the **Layout** tab, and then click on **Axis Titles** from the **Labels group.** Add both a horizontal and vertical axis title. When you are done, your graph should look like the one shown below.

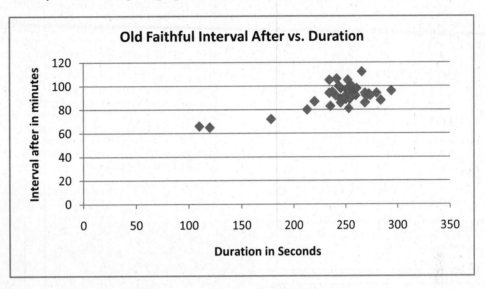

USING DDXL TO CREATE A QUICK PICTURE

If you just want a quick picture to see what the scatter plot looks like, you can produce one quickly with DDXL.

1) **Collect the data you want to work with in Excel:** Open the data set either from the CD with your book, or from the website. In a new worksheet, copy and paste the information on Duration as well as the information on Interval After. Make sure that you keep the columns exactly as they are listed in the original data.

2) **Select the columns:** Select both columns A and B by holding the left mouse down and dragging over the letters A and B on your worksheet.

3) **Access the regression feature in DDXL:** Click on the **Add-Ins** tab, and then click on **DDXL** in the **Menu Commands group.** In the drop down menu that appears, click on **Regression.** The Regression Dialog box will appear.

a. **Select the Function Type:** Click on the down arrow near **Function Type**, and click on **Correlation** from the drop down list that appears.

b. **Fill out the Dialog Box:** Make sure that the box next to **First row is variable names** is checked. Under the Names and Columns, click on Duration, and then click on the blue arrow near the x-Axis Quantitative Variable box. Click on Interval After, and click on the blue arrow near the y-Axis Variable box.   Then click on **OK.** You should see the screen shown below.

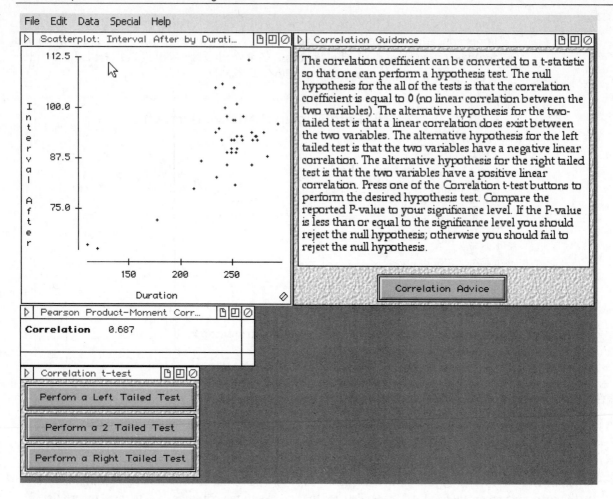

TO PRACTICE THESE SKILLS

You can apply the skills learned in this section by working on the following exercises.

1) Enter the data found in exercises 9 and 10 from Section 10-2 Basic Skills and Concepts in your textbook into Excel. Create the scatter plots, and find the linear correlation coefficients.

2) You can use Excel to work with exercises 13 through 28 from Section 10-2 Basic Skills and Concepts in your textbook. You need to enter the appropriate data given in the textbook.

3) You can use Excel to work with exercises 29 through 32 from section 10-2 Basic Skills and Concepts in your textbook. You can load the appropriate data from the CD that comes with your textbook, or find it at the website www.aw-bc.com/triola . After you have loaded the data, you should copy the appropriate columns to a new worksheet. Create the scatter plot and find the linear correlation coefficient.

SECTION 10-3 & 10-4: REGRESSION, VARIATION AND PREDICTION INTERVALS

In section 10-2 we saw that there was some evidence of linear correlation between the duration time of an eruption and the interval after the eruption. Now we want to determine this relationship in order to be able to calculate the interval time after an eruption once we know the duration of the eruption.

We have two options when working with linear regression, both of which are outlined below.

- The first option (**Add Trendline**) allows us to quickly generate the line of regression directly from our scatter plot.

- The second option uses the data analysis feature, and gives us a much more information, which will be useful in considering a more thorough analysis of the situation.

Option 1 – Add Trendline

1) **Access the Trendline feature:** Click anywhere in the **Chart** region of your scatter plot, and notice that a new **Chart Tools** tab is added to ribbon at the top of the document.   Click on the **Layout** tab, and then click on the **Trendline icon** in the **Analysis group.** Select **More Trendline Options** from the very bottom of the menu that appears.   Notice that there are options for Forecasting forward and backwards. Initially, the Trendline will automatically only include the beginning and ending input values from your data set. To extend the line to fill up more of the graph you can use the "Forecast" feature. In our example, we can go forward by 20 units, and backward by 100 units. We also want to select the options so that we can see the equation and the R-squared value on the chart.   Fill out the box as shown, and then click **Close.**

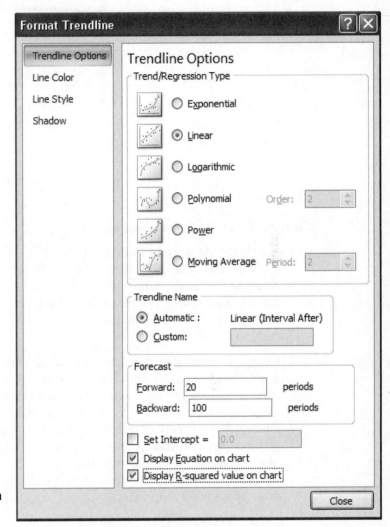

2) **Modify your initial picture:** You will probably want to reformat your font size for the equation, and reposition where the equation appears on the screen so that it does not cover any of your data points.

     a. **To modify your equation**, you can select the region containing the equation, and then right click to access the formatting features.

     b. **To move the equation**, select the region containing the equation and move your cursor in this region until a double headed arrow appears. Hold down the left click button, and move the box where you want it to be within your plot area.

     c. **Changing your Axis Values:** You may find that when you used the forecast feature, your original scaling on your scatter plot changed. You can make additional changes if desired by moving your cursor to one of the numbers on the axis and then clicking. You should see a box appear around the axis values. If you right click, you can select **Format Axis.** This will allow you to make a number of changes to your axis values.

2) **Final result:** Your picture should look similar to the one shown.

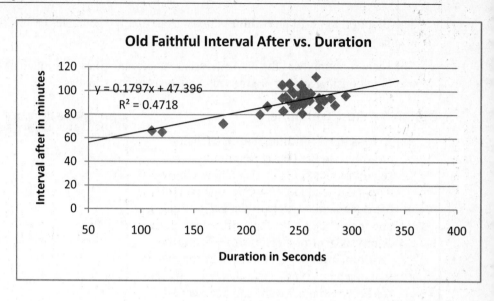

Old Faithful Interval After vs. Duration

$y = 0.1797x + 47.396$

$R^2 = 0.4718$

Option 2 – Data Analysis: Regression

1) **Access the Regression Data Analysis Tool:** Click the **Data** tab, and then click on the **Data Analysis icon** in the **Analysis group.** In the Data Analysis dialog box, scroll down until you see **Regression.** Click on this option and then click on **OK.**

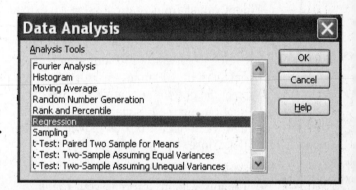

2) **Complete the Regression dialog box as shown.** The cell addresses shown assume that you entered your x value label in cell A1, and your x values (Duration times in seconds) in cells A2 through A41. Your y value label is in cell B1, and your y values (Interval After in Minutes) are in cells B2 through B41. If your data is entered in other cells, you should make appropriate adjustments.

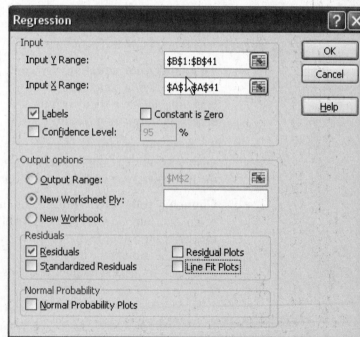

  a. **You should select both the column labels and the data in order to get the most useful, appropriately labeled output in terms of the contextual problem.** Notice that if you do include the column headings, you must check the **Labels** box. Otherwise Excel will give you an error message saying that you have selected non numeric data. You can either type in the cell addresses with a colon in between the beginning and ending cell, or you can select the cells in your worksheet. **Notice, it is imperative that you are clear** which data values represent your vertical (y) axis, and which data values represent your horizontal (x) axis.

b. **Other Options:** You can also create graphs of the Residual Plot and the Line Fit Plot directly from this Dialog Box. We leave it to you to explore some of these options more thoroughly.

3) **Click on OK**.

4) **Modify your worksheet:** You will need to resize the columns to see all the information provided clearly. To do this you can select the columns where your data is located, then in the **Home** tab, click on the down arrow by the **Format icon** in the **Cells group**. Then click on **AutoFit Column Width**. You will see the information shown below.

	A	B	C	D	E	F	G	H	I
1	SUMMARY OUTPUT								
2									
3	*Regression Statistics*								
4	Multiple R	0.686912337							
5	R Square	0.471848559							
6	Adjusted R Square	0.457949837							
7	Standard Error	7.00225424							
8	Observations	40							
9									
10	ANOVA								
11		*df*	*SS*	*MS*	*F*	*Significance F*			
12	Regression	1	1664.575551	1664.575551	33.94906058	9.8461E-07			
13	Residual	38	1863.199449	49.03156444					
14	Total	39	3527.775						
15									
16		*Coefficients*	*Standard Error*	*t Stat*	*P-value*	*Lower 95%*	*Upper 95%*	*Lower 95.0%*	*Upper 95.0%*
17	Intercept	47.39635831	7.637189976	6.205994411	2.9708E-07	31.93567562	62.85704099	31.93567562	62.85704099
18	Duration	0.179690406	0.030839761	5.826582238	9.8461E-07	0.117258573	0.242122238	0.117258573	0.242122238
19									
20									
21									
22	RESIDUAL OUTPUT								
23									
24	*Observation*	*Predicted Interval After*	*Residuals*						
25	1	90.52205571	1.477944289						
26	2	89.98298449	5.017015507						
27	3	92.31895977	-0.318959769						
28	4	91.06112693	8.938873072						
29	5	93.2174118	-3.217411798						

Interpreting This Output

Using the **Regression** option under **Data Analysis** provides you with more information than you need, but you can cut and paste the information that you need into another worksheet, or another document. To give you an idea of what the provided information represents, the major results are briefly described below:

- **Multiple R**: This is the correlation between the input variable (Duration time) and the output variable (Interval after). Since for this example there is only one input, the value given here is the correlation coefficient, r, expressing the linear relationship between the duration time and the interval after.
- **R Square**: This is also referred to as the coefficient of determination. It represents the proportion of variation in the output that can be explained by its linear relationship with the input.
- **Adjusted R Square**: The sample R Square tends to be an optimistic estimate of the fit between the model and the real population. The adjusted R Square gives a better estimate.
- **Standard Error**: This is the standard error of the estimate, and can be interpreted as the average error in predicting the output by using the regression equation.
- **Observations**: This is the number of paired data values included in the analysis.

- **ANOVA:** You do not need to understand most of the information provided in this section for this chapter, but essentially this part of the output gives more detailed information about the variation in the output that is explained by the relationship with the input. For each source of variation, the output gives degrees of freedom (df), sum of squares (SS), the F value obtained by dividing the mean square (MS) regression by the mean square residual, and the significance of F, which is the P-value associated with the obtained value of F. A fuller treatment of the Analysis of Variance (ANOVA) can be found in chapter 12.
- **Coefficients:** These are the coefficients for your regression equation. The value listed in the first row is the y intercept of the regression line, while the value listed in the second row is the slope of the line.
- **T Stat:** This refers to a test of the hypotheses that the intercept is significantly different from zero.
- **P – Value:** This is the probability associated with the obtained t statistic.
- **Lower and Upper 95%:** These are the confidence interval boundaries for both the intercept and the slope.
- **Residuals:** This table shows you the values that would be predicted for the output when using the regression equation for each input value. The second column shows the difference between the predicted value and the actual data value.

TO PRACTICE THESE SKILLS

You can practice the skills learned in this section by working on the following:

- Open one or more of the files where you saved the data from exercises in Section 10-2 of your textbook. Using this data, find the regression lines and the other data using the **Regression** option for the paired data. Think about how you can use the information created under "Residual Output" to help you see whether there is a close linear relationship between the pairs of data values for each exercise. You can also use information generated in your tables to help you answer the questions asked for these exercises in Section 10-3 Basic Skills and Exercises in your textbook.

SECTION 10–5: MULTIPLE REGRESSION

The previous sections dealt with relationships between exactly two variables. This section presents a method for analyzing relationships that involve more than two variables. A multiple regression equation expresses a linear relationship between an output or dependent variable (y) and two or more inputs or independent variables (x values).

We will work with the data from Data Set 15: Old Faithful Geyser in Appendix B in your book. We will consider our y variable as the time interval after the eruption, and our x variables as the duration time of the eruption and the height of the eruption.

	A	B	C
1	Interval After	Duration	Height
2	92	240	140
3	95	237	140
4	92	250	148
5	100	243	130
6	90	255	125
7	65	120	110
8	92	260	136
9	72	178	125
10	93	259	115
11	98	245	120
12	94	234	120
13	80	213	120
14	93	255	150
15	83	235	140
16	89	250	136
17	66	110	120
18	89	245	148
19	86	269	130
20	97	251	130

1) **Collect the data you want to work with in Excel:** Open the data set either from the CD with your book, or from the website. In a new worksheet, copy and paste the information on Interval After, Duration, and Height. Make sure that you create the columns exactly as they are shown below. You will be using the entire set of values in each of the original columns. **The values for the independent x values must be in adjacent columns**.

2) **Access the Regression Data Analysis Tool:** Click the **Data** tab, and then click on the **Data Analysis icon** in the **Analysis group**.

In the Data Analysis dialog box, scroll down until you see **Regression.** Click on this option and then click on **OK**.

3) **Complete the Regression dialog box as shown.** The cell addresses shown assume that you entered your y value label in cell A1, and your y values (Interval After) in cells A2 through A41. Your x value labels are in cells B1and C1 and your x values (Duration and Height) are in cells B2 through C41. If your data is entered in other cells, you should make appropriate adjustments.

Regression

Input
Input Y Range: `$A$1:$A$41`
Input X Range: `$B$1:$C$41`

☑ Labels   ☐ Constant is Zero
☐ Confidence Level: `95` %

Output options
☐ Output Range: `$M$2`
◉ New Worksheet Ply:
☐ New Workbook

Residuals
☑ Residuals   ☐ Residual Plots
☐ Standardized Residuals   ☐ Line Fit Plots

Normal Probability
☐ Normal Probability Plots

OK   Cancel   Help

a. **You should select both the column labels and the data in order to get the most useful, appropriately labeled output in terms of the contextual problem.** Notice that if you do include the column headings, you must check the **Labels** box. Otherwise Excel will give you an error message saying that you have selected non numeric data. You can either type in the cell addresses with a colon in between the beginning and ending cell, or you can select the cells in your worksheet. **Notice, it is imperative that you are clear which data values represent your vertical (y) axis, and which data values represent your horizontal (x) axis.**

4) **You will see the information given below**. From this information, you can create the equation: Time Interval After = 52.8 + .18 (Duration) - 0.046 (Height). For a discussion on other important elements, refer to the discussion in section 10-5 of your textbook.

	A	B	C	D	E	F	G	H	I
1	SUMMARY OUTPUT								
2									
3		Regression Statistics							
4	Multiple R	0.689797773							
5	R Square	0.475820968							
6	Adjusted R Square	0.447486966							
7	Standard Error	7.069511404							
8	Observations	40							
9									
10	ANOVA								
11		df	SS	MS	F	Significance F			
12	Regression	2	1678.589315	839.2946574	16.79328505	6.4621E-06			
13	Residual	37	1849.185685	49.97799149					
14	Total	39	3527.775						
15									
16		Coefficients	Standard Error	t Stat	P-value	Lower 95%	Upper 95%	Lower 95.0%	Upper 95.0%
17	Intercept	52.84847258	12.86329052	4.108472282	0.000211612	26.78497047	78.91197469	26.78497047	78.91197469
18	Duration	0.181206153	0.031267281	5.795392052	1.18855E-06	0.117852624	0.244559681	0.117852624	0.244559681
19	Height	-0.045782312	0.086458904	-0.52952686	0.599603129	-0.220964691	0.129400067	-0.220964691	0.129400067
20									
21									
22									
23	RESIDUAL OUTPUT								
24									
25	Observation	Predicted Interval After	Residuals						
26	1	89.92842549	2.071574511						
27	2	89.38480703	5.615192969						
28	3	91.37422852	0.625771483						
29	4	90.92986707	9.070132933						

TO PRACTICE THESE SKILLS

You can practice the skills learned in this section and in the previous section by working on exercises 13 through 16 from Section 10-5 Basic Skills and Concepts in your textbook.

SECTION 10-6: MODELING

We have used Excel to help us generate linear models for data sets. Although the **Regression** function does not give us the option to work with other types of models, we can generate scatter plots, and then add various types of Trend lines, including their equations. Your book shows various models that can be used with the TI-83/84 Plus calculator. In this section, we will use Excel to generate Quadratic and Exponential models for the data set given in Table 10-4 of your book.

	A	B
1	Coded Year	Population
2	1	
3	2	10
4	3	17
5	4	31
6	5	50
7	6	76
8	7	106
9	8	132
10	9	179
11	10	227
12	11	281

1) **Enter the data:** Enter the data for the Coded Year and the Population into a new worksheet.

2) **Create the scatter plot:** To create the scatter plot for this data, refer back to the instructions in section 10-2 of this manual, starting with step 3. Make appropriate adjustments to your chart size, font size, etc. to create a reasonable picture. Remember you can right click while your cursor is in any part of your chart to access formatting options for that particular region. Your scatter plot should appear as the one shown below:

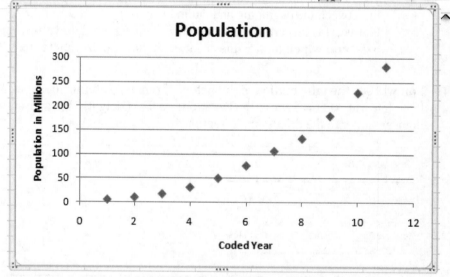

3) **Access the Trendline feature:** Click anywhere in the **Chart** region of your scatter plot, and notice that a new **Chart Tools** tab is added to ribbon at the top of the page.

    a. Click on the **Layout** tab, and then click on the **Trendline icon** in the **Analysis group.** Select **More Trendline Options** from the very bottom of the menu that appears.

    b. We will first find the Quadratic Model, so make sure that you have selected **Polynomial** and that the **Order** is set at 2.

c. We also want to select the options so that we can see the equation and the R-squared value on the chart. Fill out the box as shown, and then click **Close.** You will see your Trendline and the equation and R-squared value superimposed on your scatter plot.

4) **Make changes to your graph:** You will probably want to reformat your font size for the equation, and reposition where the equation appears on the screen so that it does not cover any of your data points. You can do this by selecting the region containing the equation, then right clicking and selecting the "Format Trendline Label" option. To move the equations, select the region containing the equation and move your cursor in this region until a crossed set of double headed arrows appears. Hold down the left click button, and move the box where you want it to be within your plot area. Your final picture should look like the one shown. Notice that with appropriate rounding, the equation produced by Excel matches the one given in your textbook.

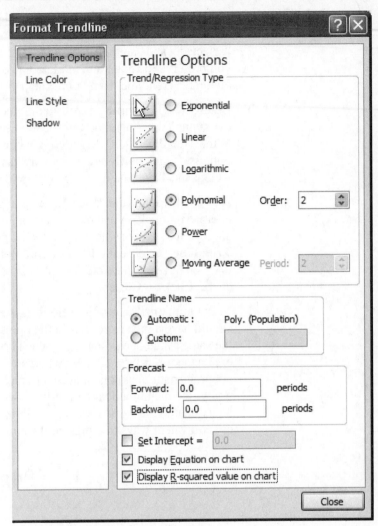

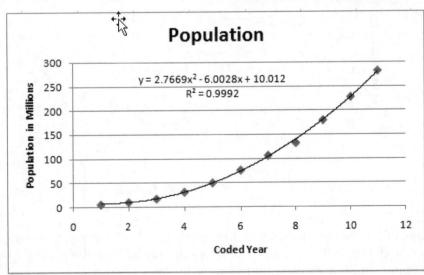

5) **Create the exponential model:** Suppose you wanted to now produce the exponential model for this same data set.

a. **Copy your Chart:** You should first select the entire Chart Area by first clicking somewhere within the white region around your actual graph. You should then see the entire Chart Area "selected", and can then copy and paste the graph to another area of your worksheet so that you preserve the quadratic model. Once you have done this, you are ready to clear the quadratic Trendline from your scatter plot, and create an exponential model.

b. **Delete the Quadratic Trendline:** Select your Chart Area. Then slowly move your cursor over your Trendline until you see the "yellow tag" appear which shows that you are pointing to the Trendline itself. Right click when you see this tag. You will see a box with several options. Click on **Delete.** You may have to do this twice to delete both your equation, as well as the Trendline itself. You should be left with your original scatter plot.

c. **Add the Exponential Trendline:** You can refer back to the instructions in step 3, or as an alternative, click on one of the data points in your scatter plot. Notice that all the data points are selected. Now right click on a data point, and in the box that appears, click on **Add Trendline.** Choose the **Exponential** option in the Add Trendline Dialog box, and make sure that you have checked off the options to display the equation and the r- squared value on the chart. Click on **Close.**

d. **Make appropriate modifications:** After making some modifications to the graph, you should have a picture similar to the one shown below. Notice that the equation is not exactly the same as the one produced by the TI-83/84 Plus calculator. The calculator uses the exponential form y = a * b^x, whereas Excel uses the form a * e^ x. You may have learned about the natural exponential function (e^x) in another math course. If you were to compute e raised to the numerical part of the power shown, you would find that the value produced is equivalent to the value for b shown in the TI-83/84 Plus screen. The two exponential models are equivalent ways to represent the equation that best fits the data.

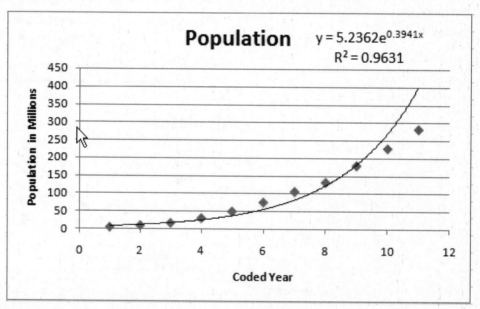

6) **Determine how close your model matches your data:** Let's suppose that you now wanted to find out how closely your model matched the data given. We will work with the Exponential model, but you can follow similar procedures for any of the models.

a. **Write down your equation:** First find the Chart Area which contains the Exponential model and equations in your Excel worksheet. We recommend writing down the equation on a piece of paper so that you don't have to keep referring to the graph.

b. **Input this equation into Excel:** You need to input this expression in cell C2 in your worksheet, assuming that your first input value (coded year) is in cell A2 and your first output value (population in millions) is in cell B2. We need to modify this equation slightly so that Excel will understand that it is a formula. Make your modifications in the formula bar at the top of the screen. You need to insert the = sign before the expression so that Excel understands that it is a formula. You need to insert a * after the coefficient to indicate multiplication. To indicate that you want to use your natural exponential function, you would type EXP(and then insert the exponent in parentheses. Your exponent needs to be adjusted by using a * after the numerical part, and then changing the x to a cell reference, in this case, cell A2. Your final formula bar should look like the one shown below. Once you have this formula entered correctly, press **Enter**. You should now see the value 7.765540 37 in cell C2.

f_x | =5.2362*EXP(0.3941*A2)

c. **Use the fill command:** You can now use the fill command to copy this formula down the rest of the column. Your table should look similar to the one below.

	A	B	C
1	Coded Year	Population	Predicted Values from Exponential Model
2	1	5	7.76554037
3	2	10	11.5166757
4	3	17	17.0797926
5	4	31	25.3301667
6	5	50	37.5658745
7	6	76	55.7120268
8	7	106	82.623657
9	8	132	122.534919
10	9	179	181.725271
11	10	227	269.507453
12	11	281	399.692717
13			

d. **Create the Quadratic Model:** You can follow similar procedures to create the predicted values for your quadratic model. The formula you would need to input for the quadratic model is shown below.

Quadratic Model: f_x | =2.7669*A2^2 - 6.0028*A2 + 10.012

TO PRACTICE THESE SKILLS

You can practice these skills by working on exercises 5 through 16 from Section 10-6 Basic Skills and Concepts in your textbook.

CHAPTER 11: MULTINOMIAL EXPERIMENTS AND CONTINGENCY TABLES

SECTION 11-1:   OVERVIEW

In previous chapters you learned that the first step in organizing and summarizing data for a single variable was to create a frequency table. Chapter 11 uses statistical methods for analyzing categorical data that can be separated into different cells. This chapter also takes a look at hypothesis tests of a claim that the observed frequency counts agree with some claimed distribution.

It is often advantageous to categorize data and create frequency counts for different variables. It is also desirable to label data according to two quantitative variables for the purpose of determining whether or not these variables are related. This data is organized by using a **contingency table** (or two-way frequency tables)**,** which consist of frequency counts arranged in a table with at least two rows and two columns. We will look at the **Chi Square test for Independence**, used to determine whether a contingency table's row variable is independent of its column variable.   You have already learned the basics for creating tables in Excel by using the **Pivot Table** wizard introduced in Chapter 4. The Pivot Table feature helps to organize, analyze and present summary data for both qualitative and quantitative variables.

CHITEST: returns the test for independence: the value from the chi-squared distribution for the statistic and the appropriate degrees of freedom.

CHITEST (actual_range, expected_range) where the actual_range is the range of data that contains observations to test against expected values and the expected_range is the range of data that contains the ratio of the product of row totals and column totals to the grand total.

SECTION 11 - 2:   MULTINOMIAL EXPERIMENTS: GOODNESS-OF-FIT

In previous chapters you looked at several different hypothesis tests that assumed that the data came from a normally distributed population. There are other less formal ways to check to see if a population is normally distributed. These might include creating a histogram and observing if the shape resembles a normal distribution. While this is not a bad approach, we will feel more secure about decisions we make if we can substantiate our findings with a formal statistical technique. A **goodness-of-fit test** is one such technique.

Excel does not contain a built in function that will perform a goodness-to-fit test. However we can use the DDXL add-in to perform a **goodness-to-fit** test.

In this section we will use the information presented in Example 2 in Section 11 – 2 of your textbook titled **World Series Games.**   You will need to enter the information from Table 11-4 into an Excel worksheet as seen here.

	A	B	C
1	Games played	Actual Contests	Expected proportions
2	4	19	0.13
3	5	21	0.25
4	6	22	0.31
5	7	37	0.31

Use a 0.05 significance level to test the claim that the actual numbers of games fit the distribution indicated by the probabilities.

1)   Enter the number of games played in column A.   In column B enter the number of actual contests.   In column C enter the expected proportions. Make sure you enter them as a decimal value.

2) **Highlight** columns A, B, and C. This is an important step so don't forget to do this. If you forget to highlight the columns which contain the number of successes and trials for both samples you will not be able to perform a hypothesis test using DDXL.

3) Click on the **Add-ins tab.**

4) Click on the **DDXL icon** in the **Menu commands** group on the left side of the **Add-ins** ribbon. Scroll down the list of options to **Tables** and click to select this option.

5) When the **Tables** dialog box opens select **Goodness-of-Fit** from Function type.

6) In the Tables dialog box:

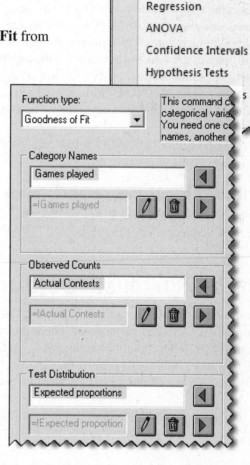

   a. **Select the column that contains the number of games played** (in this case column A) and **drag it** from the Names and Columns field to the Category Names field.

   b. **Select the column that contains the number of actual contests** (in this case column B) and **drag it** from the Names and Columns field to the Observed Counts.

   c. **Select the column that contains the expected proportions** (in this case column C) and **drag it** from the Names and Columns field to the Test Distribution.

7) Click OK.

8) DDXL returns the following test results.

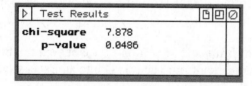

Compare these results with those found in Step 7 of the solution to Example 2 in Section 11 – 2 in your textbook.

Since the P – value is 0.0486 which is less than the significance level of 0.05 we come to the conclusion that there is sufficient evidence to reject the claim that the actual number of games in World Series contests fit the distribution indicated by the probabilities given in Table 11 – 4.

TO PRACTICE THESE SKILLS

You can practice the skills learned in this section by working through the following problems found in your textbook.

1) Work on problems 7, 9 and 13 in the Basic Skills and Concepts for Section 11-2 of your textbook.

2) Use Data Set 18 M&M.xls to work through problem 19 in the Basic Skills and Concepts for Section 11-2 of your textbook.

SECTION 11-3: CONTIGENCY TABLES: INDEPENDENCE AND HOMOGENEITY

Excel provides the tools necessary to do a chi-square test for independence, although they are not found within a single analysis tool. The process involves (1) creating a contingency table using the **Pivot table** command in Excel, (2) determining the observed frequencies, (3) determining the expected frequencies and (4) using the **CHITEST** function.

The contingency table, used to record and analyze the relationship between two or more variables, is the backbone of the chi-square test for independence in Excel.

Create a **contingency table** in Excel by entering the information presented in Table 11-6. This table represents the results from experiments with the herb Echinacea. The cells in the table represent the frequencies of those infected or not infected when given a placebo, or varying strengths of Echinacea.

	A	B	C	D
1		Placebo	Echinacea: 20% extract	Echinacea: 60% extract
2	Infected	88	48	42
3	Not Infected	15	4	10

CREATING A TABLE OF EXPECTED FREQUENCIES

The contingency table provides the actual frequencies for each cell. To perform the chi-squared test for independence we also need the expected frequencies. Excel will expect to find these expected frequencies in a separate table and not within the table you just created. We will create a second table which gives the expected frequencies for each cell.

1) Begin by determining the total for each of the rows and columns and well as the total of the totals of our original table. Compare your results with Table 11 – 7 in your textbook.

2) Copy the row headings (Infected, Not Infected) and column headings (Placebo, Echinacea:20% extract, Echinacea: 60% extract) to a different location within the same Excel worksheet. We will use this new table for the expected frequencies.

3) Find the **expected frequency** for each cell in the new table created in step (2), using the formula

$$\textbf{expected frequencies} = \frac{(row\ total)(column\ total)}{grand\ total}$$

and the appropriate cell addresses. Display

your expected frequencies correct to three decimal places

Note:

You must use relative and absolute references if you are going to copy your formula to all of your cells and not just retype the formula in each cell. For example the formula that was used to get determine the expected frequencies in cell H2 of the table below was =($E$2*B4)/$E$4. If you have forgotten how to use relative and absolute references refer back to Chapter 1 for help.

4) If all is done properly you should see

	Placebo	Echinacea: 20% extract	Echinacea: 60% extract
Infected	88.570	44.715	44.715
Not Infected	14.430	7.285	7.285

PERFORMING THE CHI-SQUARE TEST

We will test the claim presented in the **Does Echinacea Have and Effect on Colds** example found in Section 11-3 of your textbook regarding the effectiveness of Echinacea as a treatment for colds. We will use the statistical function **CHITEST** and perform the chi-square test. This function asks for the observed and expected values and will return the *p* value of the test.

1) Click on the **Formulas** tab and locate the **More Functions** icon in the **Function Library** group.

2) Click on the **More Functions** icon.

3) Highlight **Statistical** and then scroll through the list of statistical function until you highlight **CHITEST.**

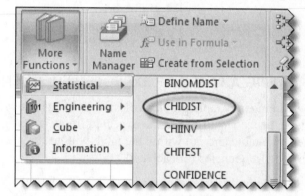

4) Click on CHITEST to open a new dialog bog.

5) When **CHITEST** dialog box opens.

   a. For the **Actual_range** highlight those cells that contain the observed frequencies (found in the first table we created). In our example we highlight the information in columns B, C and D taking care not to highlight the row and column totals.

   b. For the **Expected_range** highlight those cells that contain the expected frequencies (found in the second table we created). In our example we highlight the information in columns H, I and J.

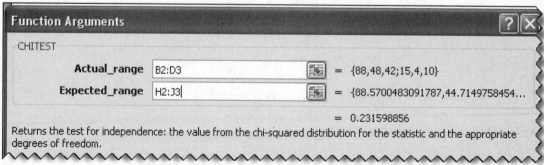

6) Click **OK.**

This returns a P-value of 0.231598. Since the P-value is greater than the significance level of 0.05 we fail to reject the null hypothesis of independence between getting an infection and the treatment received. This suggests that Echinacea is not an effective treatment for colds.

TO PRACTICE THESE SKILLS

You can practice the skills presented in this section by working through problems 5, 9 and 11 found in the Basic Skills and Concepts for Section 11-3 of your textbook.

CHAPTER 12: ANALYSIS OF VARIANCE

SECTION 12-1: OVERVIEW

In this chapter, we consider a procedure for testing the hypothesis that three or more means are equal. We will use the Analysis of Variance (ANOVA) features of Excel.

Excel tools introduced in this section are outlined below.

ANOVA SINGLE FACTOR
This feature returns summary statistics on the data, as well as Analysis of Variance information.

ANOVA: TWO FACTORS WITH REPLICATION
This feature returns summary statistics for each group in your data set, as well as Analysis of Variance information.

SECTION 12–2: ONE-WAY ANOVA

1) **Enter the data in a worksheet:** We will use the data from table 12-1 in your text book. Type the data in as shown.

	A	B	C
1	Chest Deceleration Measurements (in g) from Car Crash Tests		
2	Small Cars	Medium Cars	Large Cars
3	44	41	32
4	43	49	37
5	44	43	38
6	54	41	45
7	38	47	37
8	43	42	33
9	42	37	38
10	45	43	45
11	44		
12	50		
13			

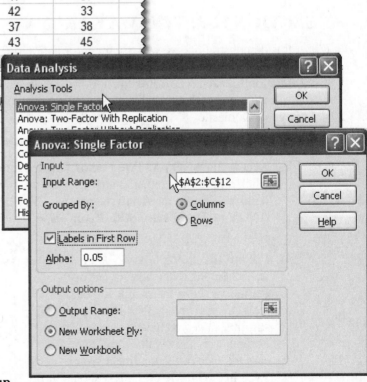

2) **Access ANOVA:** Click on the **Data** tab, and then click on the **Data Analysis icon** in the **Analysis group.** Click on **ANOVA: Single Factor**, and click on **OK.**

3) **Fill in the Dialog Box:** In the dialog box, type in "$A$2:$C$12" in the **Input Range,** or select these cells in your worksheet. Make sure that **Columns** and **Labels in First Row** are selected, that you have a check in the box by "Labels in First Row" and that **Alpha** is set at 0.05. You should have your results appear in a new worksheet. Then click on **OK.**

4) **Format the output:** You will see a table of values that is automatically selected. In the **Home** tab, click on **Format** in the **Cells group.**

Then click on **AutoFit Column Width.** You will see the table below. For a discussion of the key components of this information, read through the material presented in section 12-2 of your textbook.

Anova: Single
Factor

SUMMARY

Groups	Count	Sum	Average	Variance
Small Cars	10	447	44.7	19.34444444
Medium Cars	10	421	42.1	18.98888889
Large Cars	10	390	39	21.33333333

ANOVA

Source of Variation	SS	df	MS	F	P-value	F crit
Between Groups	162.8666667	2	81.43333333	4.094413408	0.027986345	3.354130829
Within Groups	537	27	19.88888889			
Total	699.8666667	29				

TO PRACTICE THESE SKILLS

You can practice the skills learned in this section by working on exercises 11 through 16 from Section 12-2 Basic Skills and Concepts in your textbook.

SECTION 12-3: TWO WAY ANOVA

For this demonstration, we will use the information presented in Table 12–3 of your textbook. You **MUST** enter the information for each type of treatment down a column, not across a row, and each row must begin with the category of Foreign or Domestic.

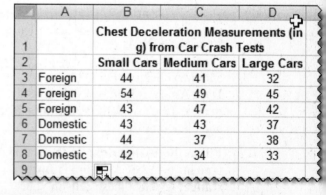

1) **Enter the data in Excel:** Set up the table as shown.

2) **Access the ANOVA: Two Factor Replication:** Click on the **Data** tab, and then click on the **Data Analysis icon** in the **Analysis group.** Click on **ANOVA: Two-Factor with Replication.** Click on **OK.**

3) **Fill out the dialog box:** Type in "$A$2:$D$8" in the box for **Input Range,** or select these cells in your worksheet. Type in 3 for **Rows per sample,** since you have 3 values for Foreign cars and 3 values for Domestic cars for each of the treatment groups. Type in 0.05 for **Alpha.** You should have

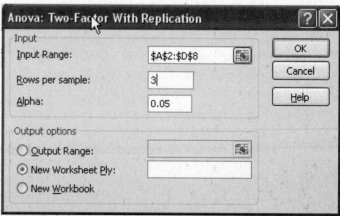

your results appear in a new worksheet. Click on **OK**.

4) **Format the results:** You will need to format the columns so that the titles all show up fully. You will see the data that is presented on the next page. For a detailed discussion of the values in this table, see section 12-3 in your textbook.

Anova: Two-Factor With Replication

SUMMARY	Small Cars	Medium Cars	Large Cars	Total
Foreign				
Count	3	3	3	9
Sum	141	137	119	397
Average	47	45.66666667	39.66666667	44.11111111
Variance	37	17.33333333	46.33333333	36.61111111
Domestic				
Count	3	3	3	9
Sum	129	114	108	351
Average	43	38	36	39
Variance	1	21	7	17
Total				
Count	6	6	6	
Sum	270	251	227	
Average	45	41.83333333	37.83333333	
Variance	20	32.96666667	25.36666667	

ANOVA

Source of Variation	SS	df	MS	F	P-value	F crit
Sample	117.5555556	1	117.5555556	5.439588689	0.037912371	4.747225336
Columns	154.7777778	2	77.38888889	3.580976864	0.060318253	3.885293835
Interaction	14.77777778	2	7.388888889	0.341902314	0.717117322	3.885293835
Within	259.3333333	12	21.61111111			
Total	546.4444444	17				

TO PRACTICE THESE SKILLS

You can practice the skills learned in this section by completing exercises 14 and 15 from Section 12-3 Basic Skills and Concepts in your textbook. Remember, when working with exercise 14, you **MUST** enter the data for each age group in columns, with male or female beginning each row.

CHAPTER 13: NONPARAMETRIC STATISTICS

SECTION 13-1: OVERVIEW

In this chapter, many of the non-parametric methods used are not immediately supported by Excel. However, we can use the DDXL Add-In.

SECTION 13-2: SIGN TEST

We can use the Add-In DDXL to work with the Sign Test.

USING DDXL TO WORK WITH SIGN TEST

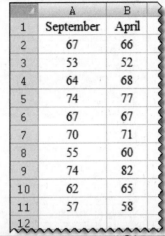

	A	B
1	September	April
2	67	66
3	53	52
4	64	68
5	74	77
6	67	67
7	70	71
8	55	60
9	74	82
10	62	65
11	57	58
12		

1) **Enter your data into Excel:** To work with the data from table 13-3 in your book, you need to enter it in Excel as shown.

2) **Select your data:** Before you access DDXL, you must select the data you will be analyzing. Select both columns A and B.

3) **Access DDXL:** Click on the **Add-Ins** tab, and then click on **DDXL** in the **Menu Commands group.** In the menu that appears, click on **Nonparametric tests.** This will open the Nonparametric Tests Dialog box.

4) **Select the Function Type:** Click on the down arrow by the Function Type box, and then click on **Paired sign test.**

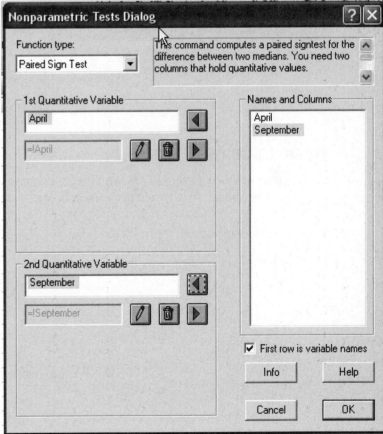

   a. **Setting the Variables:** Since you selected the entire columns, you should see the names September and April in the Names and Columns area. Click on April and then click on the Blue Arrow by the 1st Quantitative Variable box. Then click on September, and click on the Blue Arrow by the 2nd Quantitative Variable box.

   b. Click on **OK.** You will be taken to a screen like the one shown on the next page.

   c. Under **Step 1**, click on .05, since that is the significance level for the example worked out in your text book.

   d. Under **Step 2,** click on Two Tailed.

   e. Under **Step 3,** click on Compute. You should now see the results shown.

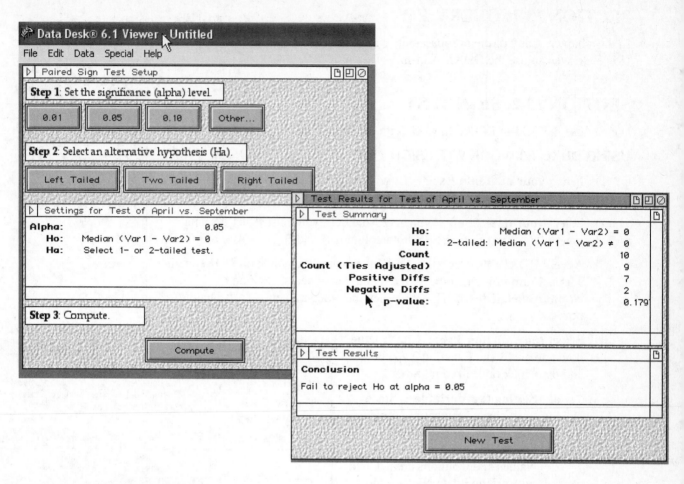

TO PRACTICE THESE SKILLS

You can practice these skills by working on exercises 9 - 12 from Section 13-2 Basic Skills and Concepts in your textbook. You would enter the data for the data in adjacent columns.

SECTION 13-3 WILCOXON SIGNED-RANKS TEST FOR MATCHED PAIRS

Again, we can use the DDXL Add-In to work with this test.

USING DDXL TO WORK WITH THE WILCOXON SIGNED RANK TEST

	A	B
1	September	April
2	67	66
3	53	52
4	64	68
5	74	77
6	67	67
7	70	71
8	55	60
9	74	82
10	62	65
11	57	58
12		

1) **Enter the data into Excel:** Make sure you have the paired data entered into Excel.

2) **Select your data:** Before you access DDXL, you must select the data you will be analyzing. Select both columns A and B.

3) **Access DDXL:** Click on the **Add-Ins** tab, and then click on **DDXL** in the **Menu Commands group.** In the menu that appears, click on **Nonparametric tests.** This will open the Nonparametric Tests Dialog box.

4) **Select the Function Type:** Click on the down arrow by the Function Type box, and then click on **Paired Wilcox.**

5) **Complete the Nonparametric Tests Dialog Box:** Since you selected the entire column, you should see the names September and April in the Names and Columns area. Click on April and then click on the Blue Arrow by the 1st Quantitative Variable box. Then click on September, and click on the Blue Arrow by the 2nd Quantitative Variable box. Click on **OK.** You will be taken to a screen like the one shown.

6) **Complete the Setup Box for the Paired Wilcoxon Signed Rank Test:**

   a. **Under Step 1:** Click on 0.05 for the significance level.

   b. **Under Step 2:** Click on Two Tailed.

   c. **Under Step 3:** Click on Compute. You will then see the screen shown.

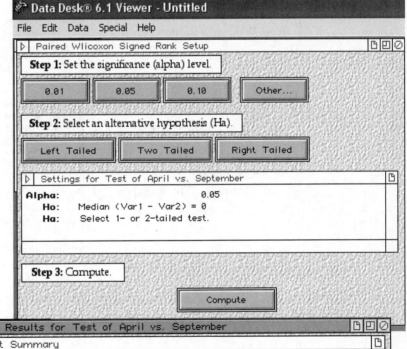

TO PRACTICE THESE SKILLS

You can practice these skills by working on exercises 5 through 12 from Section 13-3 Basic Skills and Concepts in your textbook.

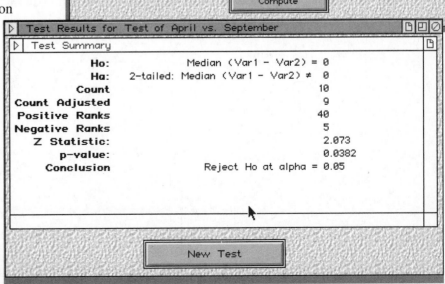

SECTION 13-4: WILCOXON RANK-SUM TEST FOR TWO INDEPENDENT SAMPLES

Although Excel is not programmed to compute the Wilcoxon Rank-Sum Test for Two Independent Samples directly, we can again use the DDXL Add-In.

USING DDXL AND THE WILCOXON RANK-SUM TEST

	A	B
1	4 Cylinders	6 Cylinders
2	136	131
3	146	129
4	139	127
5	131	146
6	137	155
7	144	122
8	133	143
9	144	133
10	129	128
11	144	146
12	130	139
13	140	136
14	135	
15		

1) **Enter the Data into Excel:** Enter the data from table 13-5 into two contiguous columns in Excel. Make sure you include column headings.

2) **Select your columns:** Make sure you select your columns before you access DDXL.

3) **Access DDXL:** You will follow the directions as indicated for the previous sections, however, you will select the **Mann-Whitney Rank Sum** option under **Function Type** in the Non-Parametric Tests Dialog box, and will see the appropriate column names for the data in table 13-5.

4) **Complete the Setup Box for the Paired Wilcoxon Signed Rank Test:**

   a. **Under Step 1:** Click on 0.05 for the significance level.

   b. **Under Step 2:** Click on Two Tailed.

   c. **Under Step 3:** Click on Compute. You will then see the screen shown.

Test Results for 4 Cylinders and 6 Cylinders

Test Summary

Alpha:	0.05
Ho:	Median (Var1 - Var2) = 0
Ha:	2-tailed: Median (Var1 - Var2) ≠ 0
Count1	13
Count2	12
Test Statistic	180.5
Z Statistic	0.626
p-value:	0.5384
Conclusion	Fail to reject Ho at alpha = 0.05

New Test

TO PRACTICE THESE SKILLS

You can practice the skills learned in this section by working with exercises 9 through 12 from section 13-4 Basic Skills and Concepts in your textbook.

SECTION 13-5: KRUSKAL-WALLIS TEST

Again, Excel is not programmed to directly perform the Kruskal-Wallis Test. As in the previous section, you can use the DDXL Add-In.

USING DDXL AND THE KRUSKAL- WALLIS TEST

1) **Enter your data into Excel:** We will use the data in Table 13-6 of your text book. To use DDXL, we must have all the values in one column, with another column containing the sample names. You must give a name for each column.

2) **Select your columns:** Make sure you select your columns before you access DDXL.

3) **Access DDXL:** You will follow the directions as indicated for the previous sections, however, you will select the **Kruskal- Wallis Test** option under **Function Type** in the **Non-Parametric Tests Dialog** box, and will see the appropriate column names for the data in table 13-5.

    a. **Select the Response Variable:** Click on **Results** in the Names and Columns box, and then click on the blue arrow by the Response Variable box.

    b. **Select the Factor Variable:** Click on **Type of Car** in the Names and Columns box, and then click on the blue arrow by the Factor Variable box.

    c. **Click on OK:** You will see the results shown.

TO PRACTICE THESE SKILLS

To practice the skills outlined in this section, you can work on exercises 5 through 12 from Section 13-5 Basic Skills and Concepts in your textbook.

	A	B
1	Type of Car	Results
2	Small Cars	44
3	Small Cars	43
4	Small Cars	44
5	Small Cars	54
6	Small Cars	38
7	Small Cars	43
8	Small Cars	42
9	Small Cars	45
10	Small Cars	44
11	Small Cars	50
12	Medium Cars	41
13	Medium Cars	49
14	Medium Cars	43
15	Medium Cars	41
16	Medium Cars	47
17	Medium Cars	42
18	Medium Cars	37
19	Medium Cars	43
20	Medium Cars	44
21	Medium Cars	34
22	Large Cars	32
23	Large Cars	37
		38
		45
		37
		33
		38
		45
		43
		42

Test Results: Type of Car by Results

T	5.774
p	0.0558
number of ties	7
T (corrected for ties)	5.835
p (corrected)	0.0541

Summary

Group	Count	Sum of Ranks	Mean Rank
Large Cars	10	109	10.9
Medium Cars	10	152.5	15.25
Small Cars	10	203.5	20.35

Boxplot: Results by Type of Car

SECTION 13-6: RANK CORRELATION

We will use Excel to help us create the test statistic for the data shown in Table 13-1 of your textbook.

USING DDXL AND THE SPEARMAN RANK TEST

1) **Enter your data into Excel:** We will use the data in Table 13-1 of your text book (Overall Quality Scores and Selectivity Ranks of National Universities). To use DDXL, we must give a name for each column.

	A	B
1	**Overall Quality**	**Selectivity Rank**
2	95	2
3	63	6
4	55	8
5	90	1
6	74	4
7	70	3
8	69	7
9	86	5
10		

2) **Select your columns:** Make sure you select your columns before you access DDXL.

3) **Access DDXL:** You will follow the directions as indicated for the previous sections, however, you will select **the Spearman Rank Test** option under **Function Type** in the **Non-Parametric Tests Dialog** box, and will see the appropriate column names for the data in table 13-1.

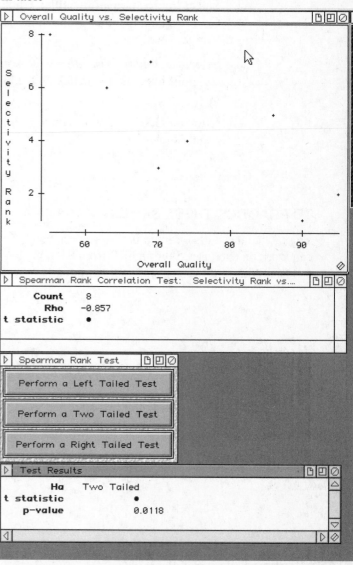

   a. **Select the x-axis Quantitative Variable:** Click on **Overall Quality** in the Names and Columns box, and then click on the blue arrow by the x-axis Quantitative Variable box.

   b. **Select the y-axis Quantitative Variable:** Click on **Selectivity Rank** in the Names and Columns box, and then click on the blue arrow by the y-axis Quantitative Variable box.

   c. **Click on OK:** You will be taken to a screen where you see the partial results, and will get to choose a Left Tailed, Two Tailed, or Right Tailed test. **Click on Two Tailed.** You will then see the results shown.

TO PRACTICE THESE SKILLS

You can practice the skills from this section by working on exercises 9 through 18 from Section 13-6 Basic Skills and Concepts in your textbook.

CHAPTER 14: STATISTICAL PROCESS CONTROL

SECTION 14–1: OVERVIEW

In this chapter, we address changing characteristics of data over time using control charts. These charts help to monitor variations by collecting data from samples at various points within the process,. If the chart indicates that the process is currently under control then it can be used with confidence to predict the future performance of the process. If a control chart indicates that the process is not in control, the pattern it reveals may help to determine the source of variation that needs to be eliminated in order to bring the process back into control.

The major features used in this section are ones that have already been introduced in earlier sections, and include:

Statistical Graphs to create a line graph from a set of data points.

Function Library to access **Average, Median,** and **STDEV.**

The new feature introduced in this section is how to use the **Callouts** under the Draw menu.

SECTION 14–2: CONTROL CHARTS FOR VARIATION AND MEAN

In this section, we will consider data arranged according to some time sequence. We will consider the information on Annual Temperatures of the Earth found in Table 14-1 in your textbook. This table can be found in the Chapter problem for this chapter. Be careful when entering the data.

1) **Enter the data for the years and the Annual Temperatures** in a new worksheet**.** It is not necessary to enter the data found in the last two columns of the table since this information can be determined using Excel.

2) **Determine** the **mean**, and the **range** for each row. Recall that this can be done by going to the **Formulas** tab and locate the **Insert Functions** icon in the **Function Library** group. Scroll through the Insert Function library to locate the appropriate function to find the mean (AVERAGE) . Follow the instructions outlined below to determine the RANGE.

The **range** is the difference between the highest and lowest value in the sample. This can be done in Excel using the MAX and MIN functions. For example, in the problem you are trying to work through – the range can be found by =MAX(B2:H2)-MIN(B2:H2).

3) To determine the **Range:**

    a. **Position** your cursor in cell M2

    b. **Enter the formula** = Max(B2:K2)-Min(B2:K2).

    c. Press **Enter**.

    d. **Copy the formula** down the rest of the column.

4) Before you go any further take some time to compare your Excel worksheet with the values shown in Table 14 – 1 in your textbook.

Control Chart for Monitoring Variation: The R Chart

Before we create the actual chart for the ranges, we need to compute the value for the Centerline, and the values for the Upper Control Limit (UCL) and the Lower Control Limit (LCL).

1) **Determine the mean for the sample ranges.** To do this, find the mean of the column containing the range values.

2) To find the upper and lower control limits, we make use of Table 14–2 in your textbook. **Locate the value** for D_3 and D_4. In this example we have our data in 10 columns, so we choose our values for D_3 and D_4 from the row for $n = 10$. We find $D_3 = 0.223$ and $D_4 = 1.777$.

3) **Determine the upper control limit** using the formula $D_3\overline{R}$. Your upper control limit (UCL) will be 0.0843.

4) **Determine the lower control limit** using the formula $D_4\overline{R}$. Your lower control limit (LCL) will be 0.6717.

5) **Create a scatterplot.** We want to plot the values for the ranges of the decades our sample on the vertical axis of our graph and we want the decades to be along the horizontal.

6) Use $\overline{R}$ (the mean of the range values) as the center line of the graph.

7) At this point you will need to determine the difference between the UCL and $\overline{R}$, and the difference between the LCL and $\overline{R}$. We want to create our graph so that there are three horizontal lines which are predominant in our scatterplot, one for the UCL, one for the LCL and one for $\overline{R}$. Work to set your horizontal gird lines so that they represent these three lines (UCL, LCL and $\overline{R}$)

8) Once you are satisfied that you accurately display the information required for the control chart, label the horizontal lines on your graph. This can be done using a textbox. Compare your final graph with the one shown below and in the solution to Example 3 in your textbook.

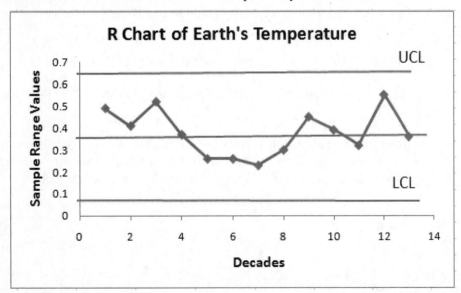

9) To add a textbox click on the **Insert** tab. Locate the **Text Box** icon in the Text Group on the right side of the ribbon.

Control Chart for Monitoring Means

You can follow the same process as the one outlined on the preceding pages to create a control chart for the sample means. A control chart of this type is used to monitor the center in a process. The centerline for this chart will represent the mean of all the sample means. We will also include a line for the upper control limit and the lower control limit.

1) **Determine the mean for the sample means.** To do this, find the mean of the column containing the mean values. This gives us the mean of the means $\overline{\overline{X}}$.

2) To find the upper and lower control limits, we make use of Table 14–2 in your textbook. **Locate the value** for A_2. Since $n = 10$ we find that $A_2 = 0.308$.

3) **Determine the upper control limit** using the formula $\overline{\overline{X}} + A_2 \overline{R}$. Your upper control limit (UCL) will be 14.135.

4) **Determine the lower control limit** using the formula $\overline{\overline{X}} - A_2 \overline{R}$. Your lower control limit (LCL) will be 13.903.

5) **Create a scatterplot.** We want to plot the values for the means of the temperatures for the decades our sample on the vertical axis of our graph and we want the decades to be along the horizontal.

6) Use $\overline{\overline{X}}$ (the mean of the mean values) as the center line of the graph.

7) At this point you will need to determine the difference between the UCL and $\overline{\overline{X}}$, and the difference between the LCL and $\overline{\overline{X}}$. We want to create our graph so that there are three horizontal lines which are predominant in our scatterplot, one for the UCL, one for the LCL and one for $\overline{\overline{X}}$. Work to set your horizontal gird lines so that they represent these three lines (UCL, LCL and $\overline{\overline{X}}$)

8) To add a textbox click on the **Insert** tab. Locate the **Text Box** icon in the Text Group on the right side of the ribbon.

9) Once you are satisfied that you accurately display the information required for the control chart, label the horizontal lines on your graph. This can be done using a textbox. Compare your final graph with the one shown on the next page and in the solution to Example 5 in your textbook.

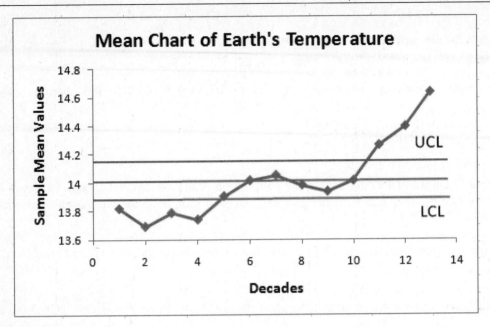

TO PRACTICE THESE SKILLS

You can apply the skills learned in this section by working on exercises 9 and 13 from Section 14-2 Basic Skills and Concepts in your textbook..

SECTION 14-3: CONTROL CHARTS FOR ATTRIBUTES

In section 14-2 we worked with quantitative data. In this section we will work with qualitative data. We will again be selecting samples of size n at regular time intervals and plot points in a sequential graph with a centerline and control limits.

We will use the information presented in Example 1 in Section 14 – 3 titled **Defective Heart Defibrillators.**

1) **Begin by entering the** number of defective defibrillators found in successive batches of 10,000 units in column A in a new Excel worksheet. These numbers can be found in Example 1.

2) **To find** $\overline{p}$

    a. **Sum** of the values in column A

    b. **Determine** the total number of defibrillators sampled. Since we sampled 20 batches of 10,000 units, the total number of defibrillators sampled is 200,000.

    c. **Divide** the sum of column A by the total number of defibrillators sampled. This gives a $\overline{p}$ value of 0.001

3) **Determine** $\overline{q} = 1 - \overline{p} = 1 - 0.001 = 0.999$

4) **Determine the upper and lower control limit** using the formula $\overline{p} \pm 3\sqrt{\dfrac{\overline{p}\overline{q}}{n}}$ where $n = 10{,}000$.

The **upper control limit** is $\overline{p} + 3\sqrt{\dfrac{\overline{p}\overline{q}}{n}} = 0.001948$ and the **lower control limit**

is $\overline{p} - 3\sqrt{\dfrac{\overline{p}\overline{q}}{n}} = 0.000052$.

5) In your Excel worksheet **determine the proportion** of the sample that each defect represents. To do this you need to divide the number of defects in each sample by the number of defibrillators in each batch.

6) **Create a scatterplot**, graphing the proportions of defects on the vertical and number of the sample along the horizontal.

7) Work to set your horizontal gird lines so that they represent these three lines (UCL, LCL and $\overline{p}$)

8) Once you are satisfied that you accurately display the information required for the control chart, label the horizontal lines on your graph. This is done using a textbox. Compare your final graph with the one shown on the next page and in the solution to Example 1 in your textbook.

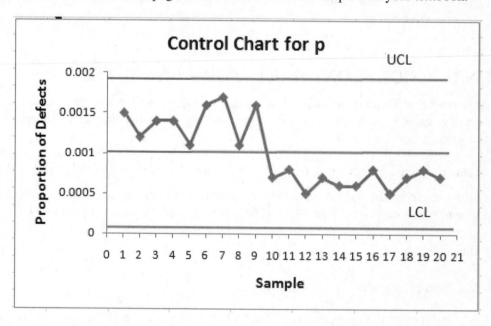

TO PRACTICE THESE SKILLS

You can apply the technology skills learned in this section by completing exercises 9, 11 and 13 from Section 14-3 Basic Skills and Concepts in your textbook.